MEDICAL
INTELLIGENCE
UNIT

Molecular Mechanisms of Cockayne Syndrome

Shamim I. Ahmad, BSc, MSc, PhD
School of Science and Technology
Nottingham Trent University
Nottingham, England

CRC Press
Taylor & Francis Group
Boca Raton London New York

CRC Press is an imprint of the
Taylor & Francis Group, an **informa** business

Molecular Mechanisms of Cockayne Syndrome

Medical Intelligence Unit

CRC Press
Taylor & Francis Group
6000 Broken Sound Parkway NW, Suite 300
Boca Raton, FL 33487-2742

First issued in hardback 2019

ISBN-13: 978-1-58706-321-3 (hbk)

Visit the Taylor & Francis Web site at
http://www.taylorandfrancis.com

and the CRC Press Web site at
http://www.crcpress.com

Library of Congress Cataloging-in-Publication Data

Molecular mechanisms of Cockayne syndrome / [edited by] Shamim I. Ahmad.
p. ; cm. -- (Medical intelligence unit)
Includes bibliographical references and index.
ISBN 978-1-58706-321-3
1. Genetic disorders. 2. Syndromes. 3. Developmental disabilities. I. Ahmad, Shamim I. II. Series: Medical intelligence unit (Unnumbered : 2003)
[DNLM: 1. Cockayne Syndrome--genetics. 2. Cockayne Syndrome--physiopathology. WE 250 M7186 2009]
RB155.5.M655 2009
616'.042--dc22

2009019070

Dedication

This book is dedicated to the sufferers of Cockayne syndrome and their parents and relations who painstakingly look after their near and dears throughout their suffering periods.

About the Editor...

SHAMIM AHMAD after obtaining his Master's degree in Botany from Patna University, Bihar, India and his PhD in Molecular Genetics from Leicester University, England, he joined Nottingham Polytechnic as Grade 1 lecturer and subsequently promoted to SL post. Nottingham Polytechnic subsequently became Nottingham Trent University where, after serving for about 35 years, he took early retirement to spend the remaining time in writing books and full time research. For more than three decades he worked on different areas of biology including thymineless death in bacteria, genetic control of nucleotide catabolism, development of anti-AIDS drugs, control of microbial infection of burns, phages of thermophilic bacteria and microbial flora of Chernobyl after nuclear accident. But his primary interest, which started 25 years ago, is DNA damage and repair, particularly near UV photolysis of biological compounds, production of reactive oxygen species and their implications on human health including skin cancer and xeroderma pigmentosum. He is also investigating photolysis of non-biological compounds such as 8-methoxypsoralen+UVA, mitomycin C, and nitrogen mustard and their importance in psoriasis treatment and in Fanconi anemia. In 2003 he received a prestigious "Asian Jewel Award" in Britain for "Excellence in Education". He is also the Editor of *Molecular Mechanisms of Fanconi Anaemia, Molecular Mechanisms of Xeroderma Pigmentosum* and *Molecular Mechanisms of Ataxia Telangiectasia* published by Landes Bioscience.

CONTENTS

Foreword xiii
Phil Hanawalt and Graciela Spivak

Preface xv
Shamim I. Ahmad

1. **Clinical Features in Cockayne and Related Syndromes** 1
Grace S. Sun and Denise W. Metry
Historical Background 1
CS Phenotypes 1
Diagnosis of CS 2
Clinical Features of CS 3
CS Genes, Complementation Groups and Phenotypes 5
Nucleotide Excision Repair in CS 6
Function of CSA 6
Function of CSB in TC-NER 6
Mechanisms of DNA Damage in CS 7
Oxidative Damage in CS 8
Proteins Involved in Apoptosis of CS Cells 8
The Cell Cycle and Chromosome Aberrations in CS 9
Chromatin Structure and CS 9
Protein-Protein Interactions with CSA and CSB 10
CS-Related Syndromes 11

2. **Transcription-Coupled Repair and Its Defect in Cockayne Syndrome** 19
Kiyoji Tanaka and Shinsuke Ito
Defective TC-NER in Cockayne Syndrome 21
Defective Transcription in Cockayne Syndrome 27
Molecular Pathogenesis of CS-Related Disorders 30

3. **Cockayne Syndrome Group B Protein and Chromatin Structure** 37
Maria Diget Aamann and Tinna Stevnsner
Biochemical Data 37
CSB and Transcription 38
DNA Repair and Chromatin 39

4. **Cell Signalling, Cell Cycle Defect and Apoptosis in Cockayne Syndrome** 43
Kiyotsugu Yoshida
Interaction of CSB with p53 44
JNK Phosphorylation of CSB 45
Functional Role for Tyrosine Phosphorylation of CRB by c-Abl 46

5. **Roles of the Cockayne Syndrome Group B Protein in Processing Oxidative DNA Damage and in Protection against Neurodegeneration** 53
Meltem Muftuoglu and Vilhelm A. Bohr
Oxidative DNA Damage 53

Base Excision Repair in Mammalian Cells 54
Cockayne Syndrome 55
Sensitivity of CSB Deficient Cells to Oxidative DNA Damaging Agents 56
Ionizing Radiation as Oxidizing Agent 56
Hydrogen Peroxide as Oxidizing Agent 56
Potassium Bromate as Oxidizing Agent 57
Paraquat as Oxidizing Agent 57
Methyl Methanesulfonate as Mutagen 57
Potential Role for CSB in BER of Oxidatively Damaged DNA 57
Mouse Models for CSB 59
Oxidative Stress, Neurodegeneration and Relations to CS Phenotype 59

6. Structural Biology of Cockayne Syndrome Proteins, Their Interactions and Insights into DNA Repair Mechanisms 65
J. Jefferson P. Perry and John A. Tainer
CSA Protein 66
CSB Protein 68
XPB Helicase 69
XPD Helicase 71
XPG Endonuclease 73
Perspectives 74

7. Cockayne Syndrome: Its Overlap with Xeroderma Pigmentosum and Other Progeroid Syndromes 81
W. Clark Lambert, Edward G. Neilan, Claude E. Gagna, George G. Kibiczak, Hon-Reen Kuo and Muriel W. Lambert
The Ultraviolet-Sensitive Syndrome (UVSS) 82
Trichothiodystrophy 83
XP/Cockayne (XP/CS) Overlap Syndrome 84
XP-B/CS Overlap Syndrome 84
XP-D/CS Overlap Syndrome 85
XP-G/CS Overlap Syndrome 86
XP-H/CS Overlap Syndrome 86
XP/CSB Overlap Syndrome 86
XP/Progeroid Overlap Syndromes 86

8. Molecular Basis and Molecular Diagnosis of Cockayne Syndrome 91
Edward G. Neilan and W. Clark Lambert
Laboratory Testing for CS 92

9. Animal and Yeast Models of Cockayne Syndrome 97
Shisheng Li
Mouse (*Mus musculus*) Models 98
Fruit Fly (*Drosophila melanogaster*) Models 101

Index 109

EDITOR

Shamim I. Ahmad
School of Science and Technology
Nottingham Trent University
Nottingham, England
Email: shamim.ahmad@ntu.ac.uk

CONTRIBUTORS

Maria Diget Aamann
Danish Centre for Molecular Gerontology and Danish Aging Research Centre
Aarhus University
Aarhus, Denmark
Email: mada@mb.au.dk
Chapter 3

Vilhelm A. Bohr
Laboratory of Molecular Gerontology
National Institute on Aging
NIH
Baltimore, Maryland, USA
Email: vbohr@nih.gov
Chapter 5

Claude E. Gagna
Department of Pathology and Laboratory Medicine
UMDNJ-New Jersey Medical School
Newark, New Jersey, USA
Email: dr.c.gagna@att.net
Chapter 7

Phil Hanawalt
Department of Biological Sciences
Stanford University
Stanford, California, USA
Email: hanawalt@stanford.edu
Foreword

Shinsuke Ito
Lineberger Comprehensive Cancer Center
Department of Biochemistry and Biophysics
University of North Carolina
Chapel Hill, North Carolina, USA
Email: shinsuke_ito@med.unc.edu
Chapter 2

George G. Kihiczak
Department of Dermatology
UMDNJ-New Jersey Medical School
Newark, New Jersey, USA
Email: kihiczak@umdnj.edu
Chapter 7

Hon-Reen Kuo
Department of Pathology and Laboratory Medicine
UMDNJ-New Jersey Medical School
Newark, New Jersey, USA
Email: kuohr@umdnj.edu
Chapter 7

Muriel W. Lambert
Department of Pathology and Laboratory Medicine
UMDNJ-New Jersey Medical School
Newark, New Jersey, USA
Email: mlambert@umdnj.edu
Chapter 7

W. Clark Lambert
Department of Pathology and Laboratory Medicine
UMDNJ-New Jersey Medical School
Newark, New Jersey, USA
Email: lamberwc@umdnj.edu
Chapters 7 and 8

Shisheng Li
Department of Comparative Biomedical Sciences
School of Veterinary Medicine
Louisiana State University
Baton Rouge, Louisiana, USA
Email: shli@vetmed.lsu.edu
Chapter 9

Denise W. Metry
Associate Professor of Dermatology and Pediatrics
Texas Children's Hospital/Baylor College of Medicine
Houston, Texas, USA
Email: dmetry@bcm.tmc.edu
Chapter 1

Meltem Muftuoglu
Laboratory of Molecular Gerontology
National Institute on Aging
NIH
Baltimore, Maryland, USA
Email: meltem.muftuoglu@bric.dk
Chapter 5

Edward G. Neilan
Children's Hospital Boston
Harvard Medical School
Boston, Massachusetts, USA
Email: edward.neilan@childrens.harvard.edu
Chapters 7and 8

J. Jefferson P. Perry
Department of Molecular Biology
Skaggs Institute for Chemical Biology
The Scripps Research Institute
La Jolla, California, USA
and
School of Biotechnology
Amrita Vishwa Vidya Peetham
Amritapuri, Kerala, India
Email: jjperry@scripps.edu
Chapter 6

Graciela Spivak
Department of Biological Sciences
Stanford University
Stanford, California, USA
Email: gspivak@stanford.edu
Foreword

Tinna Stevnsner
Danish Centre for Molecular Gerontology and Danish Aging Research Centre
Aarhus University
Aarhus, Denmark
Email: tvs@mb.au.dk
Chapter 3

Grace S. Sun
Department of Dermatology
Baylor College of Medicine
Houston, Texas, USA
Email: gssun@bcm.tmc.edu
Chapter 1

John A. Tainer
Department of Molecular Biology
Skaggs Institute for Chemical Biology
The Scripps Research Institute
La Jolla, California, USA
and
Life Sciences Division
Department of Molecular Biology
Lawrence Berkeley National Laboratory
Berkeley, California, USA
Email: jat@scripps.edu
Chapter 6

Kiyoji Tanaka
Human Cell Biology Group
Graduate School of Frontier Biosciences
Osaka University, and Solution-Oriented Research for Science and Technology (SORST)
Japan Science and Technology Agency (JST)
Osaka, Japan
Email: ktanaka@fbs.osaka-u.ac.jp
Chapter 2

Kiyotsugu Yoshida
Medical Research Institute
Tokyo Medical and Dental University
Tokyo, Japan
Email: yos.mgen@mri.tmd.ac.jp
Chapter 4

FOREWORD

We often learn the molecular mechanisms of cellular processes through the painstaking characterization of human genetic disease, beginning with the identification of a syndrome, then studying the properties of cells from victims of the disease, and ultimately cloning the genes involved so that the roles of their protein products can be studied.

The first example of a human hereditary disease with a characteristic DNA repair deficiency was the classical form of xeroderma pigmentosum, shown to have defects in early steps in the ubiquitous pathway of nucleotide excision repair (NER) and a remarkable sensitivity to sunlight-induced skin cancer. The characterization of NER and the large variety of structurally unrelated DNA lesions, which are subject to this repair pathway, greatly stimulated interest in the roles of environmental chemicals and radiations in cancer and other human diseases.

When Alfred Cockayne in 1946 first reported the disease that would be named after him, Cockayne syndrome (CS), he had no notion of the features that today fascinate numerous researchers worldwide; these include the intriguing absence of skin cancer or internal tumors in CS individuals, in spite of severe sunlight sensitivity, and the particular DNA repair defect that causes that photosensitivity, typical of most CS patients. The prominent clinical hallmarks of this devastating syndrome also include serious deficiencies in early development, growth failure and severe neurological defects.

A clairvoyant hint as to the nature of the DNA repair deficiency in CS came from the work of Lynne Mayne and Alan Lehmann, who observed that the recovery of RNA synthesis, following UV irradiation of cells from CS patients, was unusually delayed. This led them to suggest that there might be a deficiency in the repair of damage in active genes. Our understanding of the nature of the UV sensitivity in Cockayne's syndrome was then furthered by the discovery of a sub-pathway of NER, transcription-coupled repair (TCR), which is targeted to the transcribed strands of expressed genes and is defective in the cells from CS patients.

It was naturally assumed that all of the clinical features of the disease must derive from the defect in TCR. Now we know that the products of the genes primarily implicated in this disease have a complexity of functions, of which TCR is just one. For example, defects in basal transcription or in the ability to reinitiate transcription following exposure to DNA damaging agents have been documented in CS cells. Moreover, there is the possibility that the disease involves the faulty processing of oxidative DNA damage that occurs constantly in cells as a natural consequence of oxidative metabolism. In this respect the disease ties into some current models for aging that are based upon the damage inflicted by reactive oxygen species (ROS). CS patients do indeed exhibit some features of premature aging, but these are not nearly as prominent or universal as those in the progeroid syndromes, such as Werner's syndrome and Hutchinson Gilford progeria.

The existence of another rare human genetic disease, named UV sensitive syndrome (UVSS), makes the understanding of CS even more puzzling; that is because individuals with UVSS and their cells respond to UV in a manner that is indistinguishable from that of CS, and yet the affected individuals present none of the developmental and neurological deficiencies that are the hallmarks in CS patients. It has been shown that UVSS cells are proficient in the processing of oxidative base damage in DNA, a finding consistent with the hypothesis that the severe, progressive neurological and developmental problems in CS could be caused by abnormal responses to lesions induced by ROS.

This book comprehensively covers the many features of CS, from the clinical presentation to details learned from molecular studies, with provocative results that illustrate the complexity of this disease. The book should stimulate researchers and clinicians to search for further connections between the CS gene defects and consequent medical conditions. The insights gained may also be valuable for the development of new therapeutic regimens for dealing with the clinical problems raised by this rare but devastating human condition.

Phil Hanawalt, PhD
Graciela Spivak, PhD

PREFACE

Cockayne syndrome (CS) is a rare autosomal genetic disorder that was first identified almost 62 years ago by Alfred Cockayne and was named after him. The earliest publication record (PubMed) available is a paper by Marie et al in 1958.[1] Since then 815 research papers including excellent reviews have been published (PubMed, December 2008), yet we are a long way from fully understanding the exact molecular mechanisms of this disease. Ironically, like many other inborn genetic defects, CS is still incurable; the mean life expectancy of the patients is 12.5 years.[2]

Major milestones in the study of CS were the discovery that the patients have a defect in DNA repair,[3] the identification of the two complementation groups CSA and CSB,[4] and the finding that CS cells were defective in the specialized pathway of nucleotide excision repair, transcriptional-coupled repair (TCR), that removes certain lesions from actively transcribed DNA.[5]

There are numerous variations in the clinical features exhibited by sufferers of CS and related syndromes, and they are fully explored in Chapter 1 by Malcolm Taylor. Although only two genes CSA and CSB have been directly associated with CS, a large number of physical and physiological disorders can be observed in the patients. Two major features are growth retardation and neurodevelopmental impairments. Other clinical features include microcephaly, ophthalmologic abnormalities such as progressive pigmented retinopathy and retinal atrophy, photosensitive skin, rash and progeroid facies. CS patients also suffer from demylination, ataxia, cerebral atrophy and calcification of brain (especially in basal ganglia), gait defect, sensoneurial hearing loss leading to deafness and dental carries. From this it is hard to imagine that mutation in CSA or CSB gene could lead to so many clinical features; thus it is likely that more mutations leading to CS may be found. Future studies on more CS genes and the link between the mutations ultimately may help pick apart the complexity and variety of clinical manifestations in CS.

At the molecular level a major feature of CS is the impairment of TCR. TCR removes the blockage of transcription at damaged sites in DNA; otherwise the blockage can lead to apoptosis. Hence the cells from CS patients have enhanced apoptosis, and this has been hypothesized to lead to neurological deteriorations and stunted growth. The reduced proneness to skin cancer in CS has also been associated with this; after all, dead cells do not form tumors. In Chapter 2, Kyoji Tanaka and Shinsuke Ito have presented an exhaustive review of TCR and mutations in the CSA and CSB leading to CS.

Tinna Ventrup and her co-author in Chapter 3 extended the information on the roles of CSB in chromatin remodeling and highlighted the consequence of mutations in CSB patients.

Chapter 4 presents the roles of CS proteins in cell signaling, cell cycle defect and apoptosis. Importantly, in this chapter it is shown that in addition to their participation in TCR, the CS proteins play roles in base excision repair

(BER). An interesting question raised in this chapter is what kind of damage the DNA of neuron cells sustains that can lead to neurodegeneration. Most studies with CS cells have been carried out with agents that are non-biological or cannot access the central nervous system such as UV and cisplatin. In Chapter 5 Muftuoglu and Bohr address this question and suggest a role for the CSB protein in processing oxidative DNA damage and protecting the cells from neurodegeneration due to endogenously generated ROS, and also ROS from exogenous sources (e.g. ionizing radiation).

It is interesting to note that most common type of DNA damage induced by ROS is single strand DNA breaks (SSBs). An estimated frequency of this type of damage is tens of thousands per cell per day from direct attack by intracellular metabolites and from spontaneous DNA decay.[6] If this is the case one might ask what roles might TCR and BER play in removing the SSBs, as it is known that simple SSBs can be repaired by ligases, while complicated SSBs may be substrates for BER. It is likely there are more defense mechanisms to combat the ROS attack as mentioned below.

In Chapter 6 John Tainer and his co-author Jeff Perry have addressed the roles of CS proteins as can be deduced from their structures. In their studies they also included the roles of XPB and XPD helicases and XPG endonuclease in TCR. Certain mutations in XPB, XPD or XPG can cause combined CS and xeroderma pigmentosum (XP) symptoms, or XP/CS.

In Chapter 7 Lambert et al eloquently addressed various syndromes overlapping CS; these include XP, triothiodystrophy (TTD), progeroid, cerebro-oculo-facial-skeletal and UVSS syndromes. Although a number of well identified complementation groups that have combined XP and CS, due to mutation in XPB, XPD or XPG; these could be considered as a separate (although related) diseases, XP/CS.[7]

In Chapter 8, Neilon and his co-authors have addressed the difficulties related to the clinical diagnosis of CS.

The final chapter addresses "the isolation and analysis of animal and yeast models". These include mice, fruit flies, *Saccharomyces cerevisiae*, and *S. pombe*. These models have been created to have mutations in the genes homologous to human CSA, CSB and XPB or XPG.

There is no evidence that interest in CS is waning. In recent years, the role of endogenously produced ROS, particularly in mitochondria, in inducing DNA damage leading to a variety of neurodegeneration such as motor neuron disease, CS, ataxia telangiectasia, Alzheimer's, Parkinson's, amyotrophic lateral sclerosis, Huntington disease, progeroid syndrome, frontotemporal lobar degeneration and even normal aging has drawn considerable attention. The defense mechanisms and defense enzymes such as superoxide dismutase, peroxidase, and catalase, have been the focus of important studies.

The editor of this book (SIA) has considerable interest in this field; recent studies have revealed a number of new enzymes (unpublished data) that may be responsible for the scavenge of ROS. Our future studies might show if deficiency of any of these newly discovered enzymes (as a result of genetic mutations) may lead to the neurodegeneration and other ROS-induced diseases. We hope that this book will stimulate both experts and novice researchers in the field with excellent overview of the current status of research and pointers to future research goals. The insights gained may also be valuable for the development of new therapeutic regimens for dealing with the clinical problems raised by this rare but devastating human condition.

Shamim I. Ahmad, BSc, MSc, PhD

References

1. Marie J, Leveque B, Hesse JC. Nanism with deaf-mutism and retinitis pigmentosa (cockayne syndrome). Arch Fr Pediatr 1958; 15:1101-1103.
2. Nancy MA, Berry SA. Cockayne syndrome: review of 140 cases. Am J Med Genet 1992; 42:68-84.
3. Lehman AR. Cancer-associated human genetic diseases with defects in DNA repair. Cancer Res Clin Oncol 1981; 100:117-124.
4. Tanaka K, Kawai K, Kumahara Y et al. Genetic complementation groups in cockayne syndrome. Somatic Cell Genet 1981; 7:445-455.
5. Venema J, Mullenders LH, Natarajan AT et al. The genetic defects in Cockayne syndrome is associated with a defect in repair of UV-induced DNA damage in transcriptionally active DNA. Proc Natl Acad Sci 1990; 87:4707-4711.
6. Caldecott KW. Single strand break repair and genetic disease. Nature Rev Genet 2008; 9:619-631.
7. Vermeulen W, Jaeken J, Jaspers NG et al. Xeroderma pigmentosum complementation group G associated with Cockayne syndrome. Am J Hum Genet 1993; 53:185-192.

CHAPTER 1

Clinical Features in Cockayne and Related Syndromes

Grace S. Sun* and Denise W. Metry

Abstract

Cockayne syndrome is an autosomal recessive disorder in which patients develop growth retardation and neurodevelopmental impairment. The clinical symptoms may be present at birth and typically develop by age two. Patients usually do not survive to adulthood. In addition to growth and developmental delay, affected children have a typical facies, ophthalmic and auditory disorders and sun-sensitivity. The mutation of genes *CSA* or *CSB* leads to impaired DNA-repair and increased DNA damage to oxidation, radiation and UV exposure. Patients with UV-Sensitive syndrome and XP also have increased sun sensitivity but do not share the increased cellular apoptosis seen in Cockayne syndrome. The gene products of *CSA* and *CSB* have many complex protein interactions involving DNA-repair, cell cycle regulation and chromatin structure. Research continues to report more proteins interacting with CSA and CSB proteins.

Historical Background

In 1936, a London physician, Edward Alfred Cockayne, described two siblings with a syndrome of "dwarfism, retinal atrophy and deafness".[1] Since then, more than 180 cases of Cockayne syndrome (CS) have been reported in the literature. CS, an autosomal recessive disorder, has no apparent gender or ethnic predilection.[2] The classic clinical features include profound growth retardation, microcephaly, ophthalmologic abnormalities, neurodevelopmental impairment, photosensitive skin rash and progeroid facies.[3] While most CS children are of normal birth weight (Fig. 1), profound growth retardation and microcephaly become apparent within the first two years of life. This type of growth retardation affects the child's weight significantly more than the height and is termed "cachectic dwarfism".[4]

CS Phenotypes

CS has been divided into four clinical phenotypes:

CS Type I is also known as the "classic" form. Affected children are of normal birth weight, but develop profound growth retardation by age two (Fig. 1). Children are typically below the fifth percentile for height, weight and head circumference and also tend to develop progressive audiologic, visual and neurologic deficits. Death usually occurs by the first or second decade of life (Fig. 2).

CS Type II is also known as cerebro-oculo-facial syndrome (COFS), Pena-Shokeir syndrome Type II, or "conatal" CS. These children are even more severely affected than those with Type I, presenting at birth with intrauterine growth retardation, followed by poor postnatal growth and rapidly progressive neurologic impairment. Thirty percent have ophthalmologic abnormalities,

*Corresponding Author: Grace S. Sun—Baylor College of Medicine, Dept of Dermatology, 1709 Dryden Rd. Suite 1050, Houston, TX 77030, USA. Email: gssun@bcm.tmc.edu

Molecular Mechanisms of Cockayne Syndrome, edited by Shamim I. Ahmad.

Figure 1. A CS infant with normal birthweight. Bender M et al. Pediatric Dermatology 2008; 20(6)538-540.[75]

including congenital cataracts and/or early structural eye abnormalities. Many patients have arthrogryposis with contractures of the hips, knees and ankles, kyphosis or scoliosis. Death typically occurs by age 6 or 7 years.

CS Type III symptoms are milder than Type I, manifesting later and with slower disease progression.

Another variant of CS is Xeroderma Pigmentosum-Cockayne syndrome (XP-CS).[4] Affected children have the typical facial freckling and skin cancer predisposition characteristic of patients with XP, in addition to hypogonadism, mental retardation, spasticity and short stature. This type is distinguished by the absence of neurological impairment or progeroid facies.[5]

Diagnosis of CS

At present, diagnostic criteria have been established only for CS Type I. The presence of two major criteria is necessary to establish the diagnosis in an infant or young child. In older children, the presence of two major criteria and three minor criteria are required.

Major Criteria

Growth retardation—height, weight and head circumference below the fifth percentile by age two ("cachetic dwarfism").

Neurodevelopmental impairment—developmental delay, progressive mental retardation and leukodystrophy evident on magnetic resonance imaging (MRI).

Minor Criteria

Demyelinating peripheral neuropathy

Retinal pigmentation

Figure 2. The CS patient in Figure 1 during early childhood. Bender M et al. Pediatric Dermatology 2008; 20(6)538-540.[75]

Cutaneous photosensitivity
Sensorineural hearing loss
Dental caries
Progeroid facies[4,5]

Clinical Features of CS (see also Table 1)

The most common cutaneous manifestation of CS is a photosensitive skin eruption. Though classically described as a malar rash, photodermatitis can develop within any sun-exposed site and typically results in severe pruritis.[6,7] Other conditions with defects in nucleotide excision repair (NER) such as XP, trichothiodystrophy (TTD) and UV sensitive syndrome ($UV^{S}S$) also exhibit sensitivity to ultraviolet light. However, CS patients Types I-III have no increased risk of skin or visceral malignancies as do those with XP or TTD. Unlike CS patients, patients with $UV^{S}S$ have extreme sensitivity to UV light but no other manifestations.[8] Significant facial lipoatrophy, large ears and a thin nose contribute to the progeroid facies typical of CS.[4]

Progressive ocular abnormalities are also common in CS. Patients may have mitotic pupils that do not dilate appropriately in response to ophthalmic mydriatics. Enophthalmos results from periorbital lipoatrophy, causing the eyes to appear "sunken". Retinal pigmentation in a "salt and pepper" pattern is another characteristic feature.[6] As noted previously, cataracts are a feature most common to CS Type II.[5]

CS severely affects the nervous system, resulting in profound neurodevelopmental impairment. Patients typically have microcephaly, mental retardation and developmental delay. Neurological studies, including electrophysiology (electromyogram, nerve conduction), nerve biopsy, computed tomograph (CT) and MRI can reveal leukodystrophy, characterized by diffuse, extensive demyelination of the central and peripheral nervous systems.[4,5,9,10] Pericapillary calcifications form in the cerebral cortex and basal ganglia, as well as ventricular enlargement as a result of neuronal loss in the cortex and cerebellum.[9,10] Extensive peripheral nerve demyelination results in sensorineural hearing loss, muscle spasticity and atrophy, hypo- or hyper-reflexia and loss of ambulation. Neuronal loss in the cerebellum and cerebral cortex leads to seizures, tremors, ataxia and affects speech.

The skeletal manifestations of CS also contribute to the inability of many CS patients to ambulate. Arthrogryposis, manifested by kyphosis and scoliosis, gives patients a "hunched-over" appearance and flexion deformities of the hips, knees and ankles may also hinder ambulation. Other radiographic findings include sclerotic epiphyses and increased density of the calvarium.

Suggested CS Evaluation

The following studies are suggested to both establish the diagnosis and estimate disease severity:

1. Routine laboratory tests to assess hepatic and renal function and identify other potential metabolic abnormalities, including glucose and calcium levels.
2. Chromosome analysis to exclude other chromosome abnormalities.
3. Chromosomal breakage studies to exclude XP and Bloom syndrome.
4. Brain CT or MRI to identify atrophy, enlarged ventricles or other abnormalities.
5. Spinal and pelvic X-rays to evaluate for skeletal defects.[9]
6. Plotting of height, weight and head circumference.
7. Developmental assessment for delays.
8. Audiometry to assess hearing loss.
9. Electroencephalogram (EEG) to identify seizure activity.
10. Electroretinogram and ophthalmologic examination to identify cataracts, retinal depigmentation and action potential velocity.
11. Nerve conduction studies to quantify the degree of peripheral nerve demyelination.
12. Dermatologic evaluation to assess skin damage and UV-sensitivity.
13. Dental evaluation to identify dental caries or other abnormalities.[5]

Table 1. Clinical features of CS

System	Diagnostic Criteria	Other Features Present in ≥ 10% of Patients
Growth	Growth retardation "cachetic dwarfism"	
Neurological	Neurodevelopmental impairment, Leukodystrophy, demyelinating peripheral neuropathy	Spasticity, muscle atrophy, hypo/hyperreflexia, abnormal gait/ataxia, inability to walk, abnormal or absent speech, tremor, seizures
Ophthalmologic	Progressive ocular abnormalities, Retinal pigmentation (~55%)	Cataracts (~36%) Miotic pupils, photophobia, hyperopia, decreased or absent tears, microphthalmia, enophthalmos, optic nerve atrophy, strabismus, nystagmus
Dermatologic	Cutaneous photosensitivity, (~75%) Thin, dry skin and dry, sparse hair	Anhidrosis, malar rash
Auditory	Sensorineural hearing loss (~60%)	
Dental	Dental caries (86%)	Absent or hypoplastic teeth, malocclusion, delayed eruption of deciduous teeth
Endocrine		Cryptorchidism, delayed/absent sexual maturation
Other	Progeroid facies	Renal insufficiency, elevated liver function tests, hepatosplenomegaly, arthrogryposis, poor feeding during infancy

Adapted from data in Nance and Berry, 1992[4]; Neilan, 2006.[5]

14. Skin fibroblasts can be submitted to a laboratory and cultured for typical CS responses to UV and oxidative damage (optional).
15. Gene mutational analysis (on research basis only).

CS Genes, Complementation Groups and Phenotypes

There are five known complementation groups in CS: CSA, CSB, CS/XPB, CS/XPD and CS/XPG. CSA, also known as ERCC8, is found on chromosome 5q12.1 and codes for a 396 amino acid, 44kd, WD-repeat protein (a protein containing 4 or more copies of a tryptophan-aspartate repeat, which is a sequence motif encoding a structural repeat). This excision repair protein interacts with ERCC6 (another excision repair protein) as well as p44 protein, a subunit of TFIIH, which is a RNA polymerase II transcription factor. Mutations in the CSA gene are responsible for the clinical phenotype of CSI, the classic form and are present in approximately 25% of CS patients.[5] 70% of these mutations are missense, nonsense or splice-site mutations, while the rest are partial deletions.

Cells with CSA mutations are extremely sensitive to UV light; although normal cells are unable to synthesize ribonucleic acid (RNA) immediately after exposure to UV light, this ability is quickly regained. In contrast, cells with CSA mutations take much longer to regain such abilities and also lack the capability to remove DNA lesions during active transcription.[12]

CSB is caused by mutations in the ERCC6 gene, found on chromosome 10q11, which are present in approximately 75% of CS patients. ERCC6 encodes a 1493 amino acid transcription-coupled repair protein that has at least seven domains conserved in DNA and RNA helicases. This protein also ameliorates the elongation of transcription products through RNA polymerase I, II and possibly III.[13] There are more than 20 different mutations reported in the ERCC6 gene, most of which are nonsense mutations or small insertions or deletions leading to frameshift mutations. These mutations lead to an early disruption of the translation process, resulting in truncated proteins.[14,15] Currently, sequence analysis is used to detect mutations in CSA and CSB. In contrast to ERCC8 mutations, which lead to the classic CS I phenotype, mutations in ERCC6 manifest in variable phenotypic expressions. For example, Cerebro-Oculo-Facial-Skeletal syndrome or Pena-Shokeir-II syndrome, once thought to be a distinct entity, is now recognized as a form of CS Type II (Table 2).

The phenotypic expression of CS is complicated and cannot always be directly correlated with the underlying mutation. As examples, a patient with the clinical features of $UV^{S}S$, without features of CS, was found to be homozygous for null mutations in ERCC6 and two siblings with the DeSanctis-Cacchione variant of XP were found to have ERCC8 mutations.[16,17]

Patients with the CS/XP phenotype may have mutations in either ERCC6 or ERCC8 and XPB, XPD, or XPG.

Table 2. Molecular genetics of CS

Complementation Group (Phenotype)	Gene	Chromosome Locus	Protein Product
Cockayne syndrome—A (CS Type I)	*ERCC8*	5q12.1	DNA excision repair protein ERCC8
Cockayne syndrome—B (CS Type II and others)	*ERCC6*	10q11	DNA excision repair protein ERCC6

Data is compiled from the following standard references: Gene symbol from HUGO; chromosomal locus, locus name, critical region, complementation group from OMIM; protein name from Swiss-Prot. Adapted from Edward G Neilan (ref. 5).

Nucleotide Excision Repair in CS

Cellular DNA is constantly undergoing damage from both endogenous and exogenous agents. The majority of damage is recognized and repaired via two methods: Nucleotide Excision Repair (NER) and Base Excision Repair (BER). NER usually recognizes and repairs intra-strand cross-links and large adducts to the DNA molecule. BER usually recognizes minor base lesions, such as oxidation or methylation of the bases and single-strand DNA breaks.[18]

A subset of NER is Transcription Coupled Repair (TCR or TC-NER) which occurs while a specific strand of DNA is being transcribed by RNA polymerase II. DNA damage results in stalling of this RNA polymerase, allowing other NER factors to arrive and repair the damage.[19,20] There are more than 28 genes encoding NER factors, most of which are involved in both TCR and Global Genome Repair (GGR or GG-NER). There are seven complementation groups of XP each represent a different NER factor: XPA, XPB, XPC, XPD, XPE, XPF and XPG. XPE and XPC/hHR23B are NER factors involved only in GGR. XPA, XPC and replication protein A (RPA) recognize DNA damage and become binding sites for other NER factors. XPB and XPD are helicase subunits of TFIIH (a basal transcription factor) that ultimately unwind DNA for access to individual DNA strands for repair. XPG and XPF/ERCC1 are endonucleases that make single strand incisions on either side of DNA damage and remove an oligonucleotide of 28 bases. The gap is filled using the undamaged complementing strand as the template.[2]

Since different segments of the genome undergo transcription at different times and with different frequencies, GGR is necessary to repair DNA damage occurring in portions of the genome not undergoing transcription at any particular time. In GGR, XPC/hHR23B, RPA and XPA recognize DNA damage.[2]

In BER, DNA glycosylases are programmed to recognize particular variants of damaged bases. Once meeting a target, the glycosylases hydrolyze the N-glycosylic bond and endonucleases (either a DNA glycosylase or an APE1), cleave the DNA backbone and remove the deoxyribose moiety until finally the base is resynthesized.[2]

Function of CSA

As mentioned previously, CSA has been shown to interact with other transcription factors. While the exact function of CSA remains to be determined, its interactions with p44 of TFIIH, XAB2 (another transcription factor) and topoisomerase I imply an essential role in transcription.[21-23] CSA is also found complexed with DDB1, Cul4A, Roc1 and CSN (COP9 signalosome) to form a complex with ubiquitin ligase functions.[24] The ubiquitination is regulated by the CSN in response to UV radiation. The relationship between CSA and CSB is still unclear. There is evidence that CSA's translocation to the nucleus following UV, cisplatin and H_2O_2 treatment is dependent on CSB, yet CSB is not found in higher concentrations in the nucleus following DNA damage.[25] Once in the nucleus, CSA binds RNA polymerase II (Pol II) to assist in TC-NER.[24]

Function of CSB in TC-NER

The role of CSB has been studied extensively using CSB fibroblasts as well as animal and fungal models. *Escherichia coli* and *Saccharomyces cerevisiae* both have proteins homologous to CSB and are thus useful models to study the function and possible interactions of CSB. CSB knockout mice, the Chinese hamster ovary cell and cells from CS patient CS1AN demonstrate the results of CSB deficiency (Table 3). With the help of such models, the diverse functions of CSB have been better elucidated.

As previously mentioned, CSB is involved in RNA Pol I, II and III-dependent transcription. CSB is thought to ameliorate elongation when the RNA polymerase is stalled at a point of damaged DNA or blockage, or simply during a natural pause.[2] During transcription CSB does not interact directly with the DNA, but with the RNA polymerase. If the polymerase stalls, CSB allows it to backtrack along the DNA. Research with *E. coli* has suggested that CSB also has the ability to push RNA polymerase forward, either by rewinding the DNA upstream of the polymerase or pulling the DNA upstream and therefore pushing the polymerase in the direction of transcription,

Table 3. CSB models

Model	Comments
Escherichia coli	MDF protein homologous to CSB[25]
Saccharomyces cerevisiae	Rad26 protein homologous to CSB[25]
Chinese hamster ovary cell	UV61 cell line deficient in CSB homologue, decreased GGR with UV treatment[26]
$Csb^{-/-}$ knockout mice liver cells, mouse embryonic fibroblasts, embryonic stem cells	Growth retardation, neurological defects, predisposed to skin cancer[27]
CSB deficient human fibroblasts	Cell lines from CS patient CS1AN[26]
CSB deficient human primary lymphoblastoid cells	Cell lines from CS patient CS1AN[26]
CSB deficient human SV40 transformed fibroblasts	From CS1AN, decreased chromosome stability, no p53 activity[26]

assisting with elongation in the case of a natural pause.[28] When a RNA polymerase stalls over a site of DNA damage, CSB induces the RNA polymerase to backtrack along the DNA strand, thereby exposing the DNA to bind by CSB and creating a DNA loop.[29] The ATPase activity of CSB is stimulated by the dsDNA loop and the energy from ATP hydrolysis is utilized to modify the RNA polymerase—DNA interface, allowing for signaling of TCR to occur.[29] In order for TCR to occur, the RNA polymerase must be ubiquitinated by a CSA-containing complex, which requires CSB for translocation to the nucleus. Once ubiquitinated, the RNA polymerase may be removed and CSB, remaining on the DNA, recruits NER factors.

CS patients, lacking CSA or CSB, have serious difficulties with TCR. Without CSA, a RNA polymerase stalled at a DNA lesion cannot be ubiquitinated and removed; therefore, TCR cannot occur. In the absence of CSB, a stalled RNA polymerase even at a natural pause will not backtrack or be pushed forward. If stalled at a DNA lesion, the inability of the RNA polymerase to backtrack will not stimulate CSB to modify the RNA polymerase-DNA interface and signal NER factors. The RNA polymerase remains stalled and cannot be ubiquitinated because the CSA complex would not be translocated to the nucleus. The final result of a permanently stalled RNA polymerase is cellular apoptosis.

Mechanisms of DNA Damage in CS

UV light is the most common DNA damaging agent in CS cells. UV exposure of DNA results in the production of cyclobutane pyrimidine dimers (CPD), which are repaired by TCR, as well as pyrimidine 6-4 pyrimidone photoproducts, which are removed by GGR. 4-nitroquinolin-1-oxide (4-NQO) causes alkali-labile ssDNA breaks and bulky adducts to DNA.[30] N-acetoxy-2-acetylam inoflurone(NA-AAF) produces N-(deoxyguanosine-8-yl)-2acetylaminofluorene (dG-C8-AAF) which is usually repaired by GGR.[31,32] Ionizing radiation causes ssDNA breaks, dsDNA breaks and oxidative base damage.[33,34] Exposure of CS cells to such agents results in decreased cell life compared to wild-type cells.

Utilizing CS models and DNA damaging agents, researchers have been able to simulate the effects of UV and oxidative damage on CS patients in vivo. Chinese hamster UV61 cells, *csb* knockout mice cells and CS patient fibroblasts are sensitive to the UV light.[26,27,32] Transfection of cDNA for CSB into CS1AN and UV61 cells decreases UV sensitivity to levels equivalent to wild-type cells.[26,35,36] CSB deficient cells were found to have decreased recovery of RNA synthesis following UV exposure compared to wild-type cells.[26,27] In addition to UV exposure, CS1AN cells exposed to NA-AAF and UV61 cells exposed to 4-NQO also experienced decreased recovery of RNA synthesis.[35,37]

Even without exposure to DNA damaging agents, there are inherent differences between CSB-deficient cells and wild-type cells. CSB deficient fibroblasts and lymphoblasts have a 50% decrease in transcription compared to normal cells, which is likely due to deficient elongation.[38] CSB deficient cells also have decreased rRNA synthesis rates and hence decreased growth rates when compared to wild-type cells.[39] Moreover, these CSB-deficient cells increase rRNA synthesis rates when transfected with cDNA with *CSB*.[39] This effect is likely due to impairment of RNA polymerase I function in the absence of CSB.[39]

Oxidative Damage in CS

Recently, D'Errico et al have shown a distinct role of CSA. Potassium bromate exposure causes oxidative cellular damage forming 8-hydroxyguanine (8-OH-Gua) and (5'S)-8,5'-cyclo 2'-deoxy-adenosine. CSA-deficient human keratinocytes and fibroblasts were hypersensitive to the potassium bromate; however, following transfection with wild-type CSA genes, the cells had decreased levels of 8-OH-Gua and increased repair rates. In in vitro studies, cells with functioning CSA but defective CSB would cleave 8-OH-Gua, while cells with defective CSA but functioning CSB would not.[40] Therefore, CSA has a role in repairing oxidative damage that is separate from CSB.

While the role of CSA is a relatively recent discovery, the deficiency of CSB has been recognized to increase the accumulation of oxidative damage for some time. Stevnsner et al found that $csb^{-/-}$ mouse liver cells and CS1AN cells had decreased repair of oxidative damage in mitochondria.[41] Tuo et al reported accumulation of oxidative purine modifications 8-hydroxy-7,8-dihydroguanine (8-oxoG)[33] and 8-hydroxy-7,8-dihydroadenine (8-oxoA).[42] They also found that in terms of endogenous oxidative damage, there was no difference between CSB-deficient and wild-type cells. However, cells deficient in both CSB and OGG1, a DNA glycosylase responsible for removing 8-oxoA:C and 8-oxoG:C in mammals, had increased amounts of endogenous oxidative damage.[43,44] CSB-deficient cells have decreased levels of OGG1 mRNA and it has been suggested that oxidative stress activates CSB, causing it to promote transcription of some of the DNA glycosylases for GG-BER.[2] Mice without OGG1 accumulate 8-oxoG and have higher spontaneous mutation rates. Comparison of $ogg1^{-/-}$, $csb^{-/-}$ mice to $ogg1^{-/-}$ mice revealed a higher spontaneous mutation rate in the mice lacking both OGG1 and CSB, suggesting that CSB plays a role in preventing spontaneous oxidative mutations in the genome.[45]

The inability to repair oxidative damage, coupled with the accumulation of oxidative products results in the severe neurodegeneration characteristic of CS. In a study of autopsied CS patients, oxidative products were collected in the globus pallidus, hippocampal dentate neurons and cerebellar dentate neurons in the form of psuedocalcified foci, foamy spheroids and free-floating material in neuropils. There was also altered expression of the glial glutamate transporters EAAT1 and GLT-1 in the globus pallidus as well as the cerebellar cortex. Proper functioning of glutamate transporters is necessary to prevent glutamate neurotoxicity.[46] Thus, a combination of excessive oxidative products and possible neurotoxicity may be the source of the neurodegeneration seen in CS patients.

Proteins Involved in Apoptosis of CS Cells

As mentioned earlier, CSB is responsible for initiating both TC-NER as well as ubiquitination of stalled RNA polymerases. Stalled RNA polymerases that cannot be removed due to the lack of ubiquitination lead to apoptosis of the cell.[47] UV or cisplatin treatment inhibits RNA synthesis which leads to increased p53 (a tumor suppressor protein) activity and consequently apoptosis.[48] In CSB-deficient cells, p53 is induced at lower doses of UV and the induction lasts longer than in wild-type cells leading to more apoptosis in CSB-deficient cells compared to wild-type cells.[49]

p53 is not the only factor determining apoptosis, as Spivak et al found that UV61 cells that have mutated p53 still have increased apoptosis following UV exposure.[50] p53 is responsible for inducing proliferating cell nuclear antigen (PCNA), a protein that forms an insoluble complex in the nuclear substructure of nondividing cells following UV irradiation. High levels of PCNA in the absence of p53 leads to DNA replication, high levels of both PCNA and p53 leads to DNA repair and low levels of PCNA or nonfunctional PCNA lead to cellular apoptosis.[51] In one study, UV61 cells produced PCNA at rates two times slower than in wild-type cells following UV exposure; this effect was negated by transfecting the UV61 cells with *hCSB*.[52] Similar results were found in human CSB-deficient cells in response to UV damage and also in response to H_2O_2-induced oxidative damage.[53]

Another important protein induced by p53 and interacting closely with PCNA, is p21, a proliferation inhibitor. CSB-deficient cells have reduced p21 turnover. An overabundance of p21 greatly prohibits cell growth, reduces p21-PCNA repair of DNA and increases the generation of oxidative products. It has been suggested that an excess of p21 is related to deficient ubiquitination of the CSA-complex.[54]

The Cell Cycle and Chromosome Aberrations in CS

Depending on the degree of UV damage, apoptosis occurs at different phases of the cell cycle. In a study by Palliti et al, UV61 cells exposed to low doses of UV induced apoptosis after passing through the S phase. However, when exposed to high doses of UV, the cells remained arrested in the G1 (growth 1) phase and underwent apoptosis before entering the S phase.[55] Another study by Proietti et al found that normal cells treated with α-amanitin (an inhibitor of RNA polymerase II) had rates of apoptosis similar to that found in UV61 cells, implying that the factor causing apoptosis was the lack of TCR.[56]

Apoptosis that occurs in CS is also associated with chromosomal aberrations. Proietti et al also found that after exposure to UV, the amount of chromosomal aberrations and sister chromatid exchanges were increased in UV61 cells compared to wild-type cells. Using flow cytometry, chromosomal aberrations were found to precede and possibly determine the likelihood of apoptosis following low levels of UV treatment.[56,57] As mentioned earlier, with high levels of UV exposure, the cells never pass into S phase and the genetic material does not condense into chromosomes.[57]

Chromatin Structure and CS

CSB was first suggested to play a role in chromatin remodeling in *Drosophila* ISWI. CSB is part of a family of DNA-dependent ATPases involved in chromatin remodeling, known as the SNF-2 like family. Citterio et al discovered that CSB has the ability to remodel mononucleosomes and chromatin of naked, singly-nicked plasmid by binding to the N-terminal tails of core histones to cause negative supercoiling.[58] Similar findings were demonstrated in the CSB model of *S. cervisiae* and CSB homologue rad26.[59]

Fousteri et al proposed an explanation for the decreased GGR seen in cells from CS patients. CS patients have decreased transcription compared to cells from nonCS patients. In studies with actinomycin D inhibiting transcription in normal cells, the cells had decreased ubiquitination of histone H2B. Ubiquitinated H2B inhibits nucleosome refolding and allows chromatin to be relaxed and open to GGR.[60,61] Also, the acetylation of histones decreases their DNA binding efficiency and makes the chromatin structure more conducive for repair. However, these histone acetyltransferases localize with RNA polymerase II and thus, when transcription in decreased, so is the quantity of histone acetyltransferases, leading to a more compact chromatin structure.[60]

CSB also affects chromatin structure indirectly via some of the genes it regulates. Histone deacetylase (HDAC) inhibitors, DNA methylation inhibitors, poly(ADP-ribose)-polymerase (PARP-1) and other chromatin disrupting factors regulate some of the same genes that CSB does. Studies show that treating CSB-cells with HDAC inhibitors allow the cells to respond to DNA damage more like wild-type cells do.[62] These factors work together to maintain heterochromatin and chromosome stability and though there are some overlaps in function, the effects of CSB-deficiency cannot be fully compensated by other factors.

Table 4. Proteins interacting with CSA

Protein	Function
p44 subunit of TFIIH	Transcription by pol I and II, NER[21]
XAB2	TCR and transcription[22]
N-terminal of Topoisomerase I	Assists in unwinding dsDNA by cutting one strand and later reannealing[23]
DDB1	Ubiquitin ligase complex component, responds to UV[24]
Cul4A	Ubiquitin ligase complex component, responds to UV[24]
Roc1	Ubiquitin ligase complex component, responds to UV[24]
COP9 signalosome (CSN)	Ubiquitin ligase complex component, responds to UV[24]
RNA polymerase II (Pol II)	Nucleotide excision repair[24]
CSB	TCR[2]

Protein-Protein Interactions with CSA and CSB

As discussed earlier, CSA is predominantly involved in transcription and in a complex with ubiquitin ligase functions (Table 4) and CSB interacts and complexes with several different proteins (Table 5).

CSB interacts with multiple subunits of TFIIH—XPB, XPD, p44 and p62. XPB is a 3'→5' helicase and part of TFIIH which unwinds DNA so that transcription or NER can take place. Mutations in XPB result in XPB/CS or TTD.[2] XPD is a 5'→3' helicase and also a subunit of TFIIH which serves the same function as XPB. Mutations in XPD result in XP, TTD and XPD/

Table 5. Proteins interacting with CSB

Protein	Function
CSA	TCR[21]
XPB and XPD of TFIIH	Transcription by pol I and pol II, NER, helicase activity[64,69]
p44 and p62 subunits of TFIIH	Transcription by pol I and pol II, NER[64]
XAB2	TCR and transcription[22]
XPG	NER, stimulation of BER[64,65]
XPA	NER[69]
RNA polymerase II (Pol II)	Transcription of mRNA[25,64]
RNA polymerase I (Pol I)	Transcription of rRNA[64]
RNA polymerase III (Pol III)	Transcription of rRNA[67]
P53	Cell cycle control, G1-arrest, NER, apoptosis[67]
Histone 2A, 2B, 3,4	Chromatin component[58]
P34 of TFIIE	Transcription by RNA Pol II[69]
OGG1	BER of 8-oxo-G and 8-oxo-A[70]
PARP-1	Maintain heterochromatin, repair oxidative damage[68]

Adapted from data in Licht et al.[2]

CS.[2] p44 of TFIIH controls the helicase activity of XPD. The precise function of p62 is more obscure, but together the 9 subunits of TFIIH are responsible for transcription and NER.

XPG is another gene that when mutated can cause XP/CS. XPG is an endonuclease that makes an incision 3′ to a DNA lesion during NER. It also can stimulate DNA glycosylase Nth1, which is involved in BER of oxidative damage.[63] CSB and XPG are found together in a complex and have been found to interact in vitro.[64,65]

The *XPA* gene product is known to assist with the early stages of NER. It is suggested that it may recognize DNA lesions and align proteins in GG-NER and TC-NER.[66]

The relationship between CS and RNA polymerase III is more implied than concrete. During metaphase, CSB-deficient cells have unstable 5S rRNA suggesting that CSB contributes to Pol III functions.[67]

PARP-1 is responsible for certain gene induction, heterochromatin maintenance, chromosome stability and also for repair of oxidative damage to the genome. PARP-1 binds to ssDNA breaks and becomes activated to add polymers of ADP-ribose to proteins. One of these proteins is CSB, which loses its DNA-dependent ATPase activity after attachment of the polymer.[68]

The interactions between CSB and OGG1, Pol I, Pol II, p53 and the N-terminal tails of histones have been discussed earlier.

CS-Related Syndromes (See Tables 6, 7)

UV Sensitive Syndrome

Patients suffering from UV^SS are sensitive to UV light and burn easily, resulting in telangiectasias and freckles, but unlike patients with XP, there is no predisposition to skin cancer development. Also unlike CS, UV^SS patients have no neurologic, developmental or growth impairment.

In UV^SS the CSB gene, *ERCC6*, is mutated resulting in total loss of gene product. CS patients have a truncated and nonfunctional CSB protein, which results in more clinical manifestations than simply the absence of the protein. This suggests that a mutated CSB protein prevents important

Table 6. Gene defects of CS-related disorders

Disorder	Gene Defect	Comments
Cockayne syndrome	*CSA, CSB*	
UV-Sensitive syndrome	*CSB (ERCC6)*	
Xeroderma Pigmentosum A	*XPA*	Increased rate of internal tumors[73]
Xeroderma Pigmentosum B	*XPB*	
Xeroderma Pigmentosum C	*XPC*	No neurological symptoms, not involved in TCR[73]
Xeroderma Pigmentosum D	*XPD, ERCC6*	De Sanctis-Cacchione syndrome is a XPD variant
Xeroderma Pigmentosum E	*DDB2*	Mild skin manifestations, no neurological symptoms, late tumor onset
Xeroderma Pigmentosum F	*ERCC4*	Mild symptoms
Xeroderma Pigmentosum G	*RAD2, ERCC5*	Variable manifestation
Xeroderma Pigmentosum Variant (XPV)	*PolH*	Gene product is polymerase-eta
Trichothiodystrophy	*XPB, XPD*	
CS/XP	*CSB* plus *XPB/XPD/XPG*	

Table 7. Clinical features of CS, UV^sS, XP, XP/CS, TTD

Clinical Feature	CS	UV^sS	XP	XP with ND[a]	CS/XP	TTD
Skin						
Photosensitivity	+	+	+	+	+	+/–
Increased pigmentation	+/–	+	+	+	+	–
UV-induced skin tumors	–	–	+	+	+	–
Eyes						
Photophobia	+	–	+	+	+	+/–
Congenital cataracts	+	–	–	–	–	+
Pigmentary retinal degeneration	+	–	–	–	+	–
Somatic						
Skeletal defects	+	–	–	–	unk	+
Short stature	+	–	–	+/–	+	+
Immature sexual development	+	–	+/–	+/–	+	+/–
Internal tumors	–	–	+	+	unk	–
Nervous System						
Developmental delay	+	–	–	+	+	+
Sensorineural deafness	+	–	–	+	+	–
Progressive neurological degeneration	+	–	–	+	+	unk
Primary neuronal degeneration	–	–	–	+	–	–
Demyelination	+	–	–	–	+	+
Cerebral atrophy	+	–	–	+	+	+/–
Cerebellar atrophy	+	–	–	+	+	–
Calcification of basal ganglia	+	–	–	–	+	+/–
Adapted from Spivak 2005[8] and Kraemer 2007[72]						

[a]ND: neurological disease.
unk: unknown

protein-protein or protein-DNA interactions. UV^sS patients, similar to CS patients, have impaired TC-NER, decreased recovery of RNA and DNA synthesis and increased p53 accumulation after UV exposure when compared to normal cells. However, GGR of CPDs and 6-4 photoproducts from UV damage are repaired at the same rates as wild-type cells. Unlike CSB-deficient cells, endogenously-induced oxidative damaged is repaired in UV^sS cells at normal levels.[71] One possible explanation is that only 5% of oxidative lesions actually block DNA transcription and the remaining 95% of oxidative lesions do not need to be repaired for transcription to continue. Non-functional CSB proteins could permanently block transcription and result in apoptosis. This suggests that perhaps there are more mutations in UV^sS cells, though UV^sS patients do not show increased rate of cancer compared to the normal population.[8]

Xeroderma pigmentosum

XP patients present with extreme photosensitivity, photophobia, dry skin, pigmentary changes, freckling before age two, decreased skin elasticity, chronic conjunctivitis and keratitis, pingueculae (bulbar conjunctival growths), pterygia (corneal conjunctival growths) and basal cell and squamous cell carcinomas and melanomas before age 10 years (Fig. 3).[72,73] Approximately 20% of CS patients also have neurological deficits; CS/XP patients have cutaneous features of XP and progressive neurological degeneration.[74]

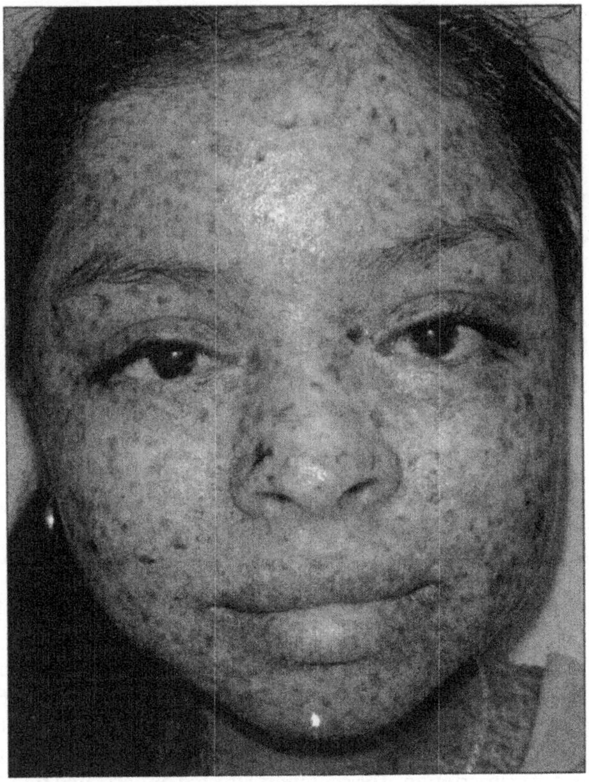

Figure 3. Child with a mutated gene for XPC.

There are eight XP complementation groups and studies show that all except XPV (XP-variant) are involved in NER. XPA recognizes DNA damage and aligns NER proteins allowing XPF and XPG nucleases to excise the damage. XPB and XPD are helicase subunits of TFIIH. XPC recognizes DNA lesions to be repaired by GGR and is not involved in TCR. XPE interacts with transcription factors, ubiquitin complexes and chromatin (affected patients have a relatively mild XP phenotype). XPF is a nuclease cutting on the 5′ side of the DNA lesion. XPF knockout mice (and one XPF patient) have kidney and liver failure, growth retardation and premature death.[74] However, most XPF patients have only one severely mutated allele and thus only a mild expression of XP.[73] XPG patients range from mild to severe manifestations of CS/XPG. XPV patients have a mutation in polymerase η that is responsible for sloppy, single-base error-proof repair that allows transcription to continue.

Trichothiodystrophy

TTD patients are known for sulfur-deficient brittle hair that is unique for its alternating light and dark banding pattern, often referred to as a "tiger tail," seen under light microscopy (Fig. 4). The clinical spectrum ranges from solely hair involvement to small stature, mental retardation, ichthyosis and an unusual facies, with or without photosensitivity (Fig. 5).[73] Unlike XP patients however, those with TTD do not have pigmentary changes or increased susceptibility to skin cancer. More severe cases may also have congenital cataracts, susceptibility to multiple infections, peripheral osteopenia and central osteosclerosis.[72] Some TTD patients have overlapping features with CS, including microcephaly, ataxia, decreased myelination of the cerebral cortex and calcification of the

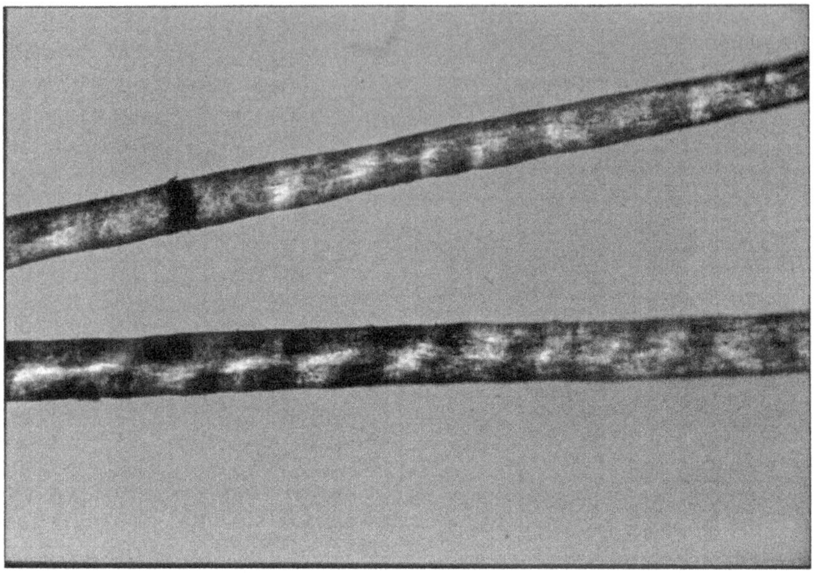

Figure 4. TTD hair under light microscopy with "tiger's tail" light and dark banding pattern.

basal ganglia.[72] PIBIDS is one variant of TTD, which is an acronym that stands for photosensitivity, icthyosis, brittle hair, infertility, decreased intelligence and short stature (Fig. 6).[72]

TTD patients have mutations in either *XPD*, *XPB* or another unknown gene. XPB is essential to initiating transcription while XPD ameliorates transcription and mutations in *XPD* simply decrease the efficiency of transcription. The cells most affected by decreased transcription efficiency

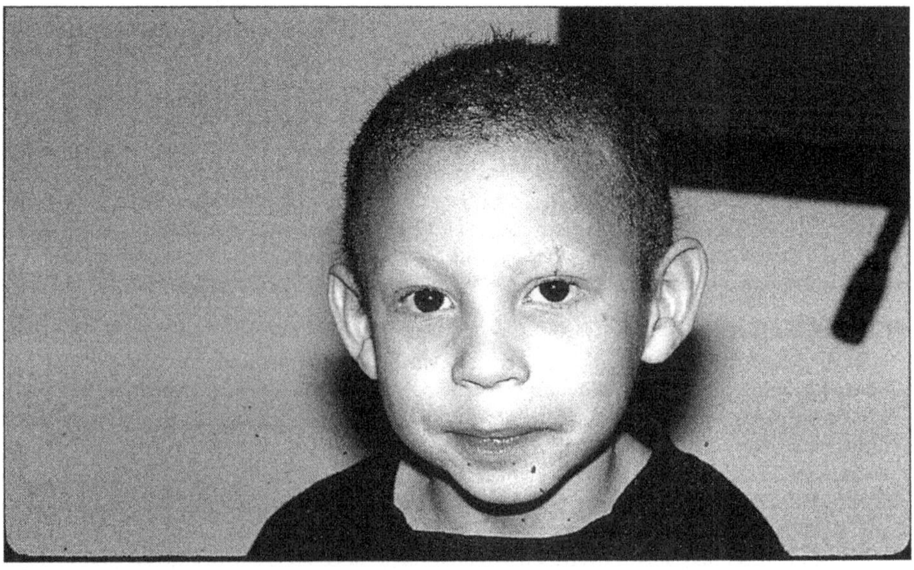

Figure 5. A patient with TTD.

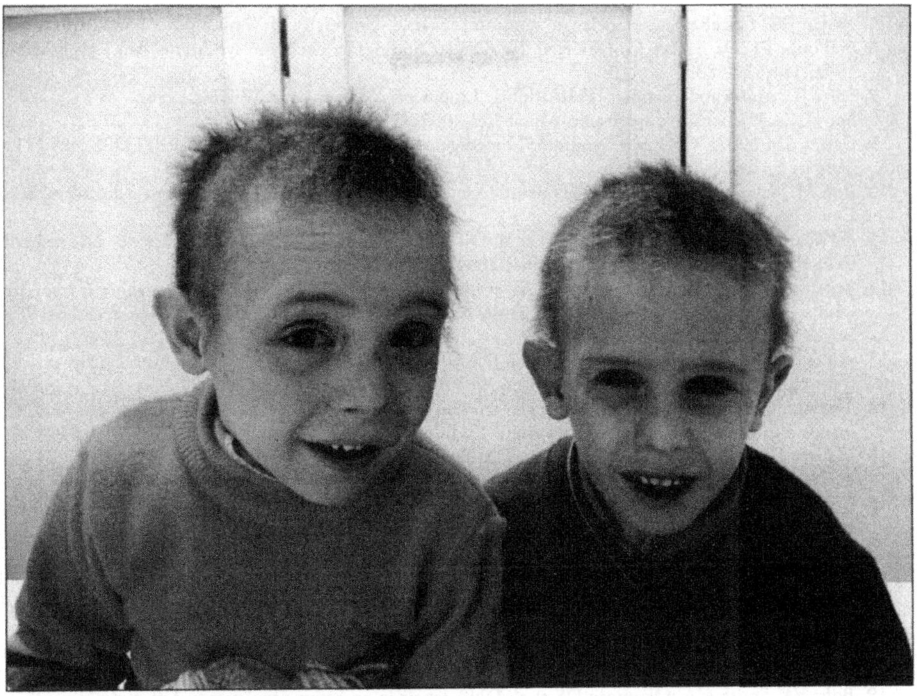

Figure 6. Two siblings with PIBIDS.

are those needed to produce large quantities of protein product: mainly matrix proteins in hair and β-globulin in erythrocytes.[73]

Prognosis in CS

Once the diagnosis of CS has been established, patients generally require multi-disciplinary care including, but not limited to, genetics, neurology, pediatrics, dentistry, ophthalmology, dermatology, audiology and physical therapy. The potential for renal compromise and hypertension needs to be assessed regularly. Physical therapy is essential to prevent joint contractures. Strict sun protective measures are also vital. Patients may maintain a normal diet and only in rare cases of malnutrition, require placement of a gastrostomy tube.

While no cure yet exists for CS, there are steps that can be taken to prevent or arrive at the diagnosis earlier, through genetic counseling. Because CS is inherited in an autosomal recessive fashion, parents of a CS patient are obligate carriers and thus have a 25% risk of having an affected child with each subsequent pregnancy. If the exact mutation(s) is identified in the child, the parents can choose to undergo prenatal testing for future pregnancies or preimplantation genetic diagnosis in the case of in vitro fertilization (IVF). Younger siblings of the proband can also be tested for CS at an earlier age and all siblings can be tested for carrier status.[5] Clinical trials are ongoing to elucidate further the relationship between CS phenotypes and specific genetic mutations.

References

1. Cockayne EA. Dwarfism with retinal atrophy and deafness. Arch Dis Child 1936; 11:1-8.
2. Licht CL, Stevnsner T, Bohr VA. Cockayne syndrome group B-cellular and biochemical functions. Am J Hum Genet 2003; 73:1217-1239.
3. Ozdirim E, Topcu M, Ozon A et al. Cockayne syndrome: review of 25 cases. Pediatr Neurol 1996; 15:312-316.
4. Nance MA, Berry SA. Cockayne syndrome: review of 140 cases. Am J Med Genet 1992; 42:68-84.

5. Neilan EG. GeneReviews: Cockayne syndrome. Seattle: University of Washington, 2006:1993-2007.
6. Traboulsi EI, De Becker I, Maumenee IH. Ocular findings in cockayne syndrome. Am J Ophthalmol 1992; 114:579-583.
7. Spitz JL, ed. Genodermatoses. Philadelphia: Lippincott Williams and Wilkins 1996.
8. Spivak, G. UV-sensitive syndrome. Mutat Res 2005; 577:162-169.
9. Carter, SM 2007, "Cockayne Syndrome", Emedicine.com. http://emedicine.com/ped/TOPIC424.HTM (August 5, 2007)
10. Itoh M, Hayashi M, Shioda K et al. Neurodegeneration in hereditary nucleotide repair disorders. Brain Dev 1999; 21:326-333.
11. de Waard H, de Wit J, Andressoo JO et al. Different effects of CSA and CSB deficiency on sensitivity to oxidative DNA damage. Mol Cell Biol 2004; 24:7941-7948.
12. Venema J, Mullenders LH, Natarajan AT et al. The genetic defect in cockayne syndrome is associated with a defect in repair of UV-induced DNA damage in transcriptionally active DNA. Proc Natl Acad Sci USA 1990; 87:4707-4711.
13. van den Boom V, Citterio E, Hoogstraten D et al. DNA damage stabilizes interaction of CSB with the transcription elongation machinery. J Cell Biol 2004; 166:27-36.
14. Troelstra C, Hesen W, Bootsma D et al. Structure and expression of the excision repair gene ERCC6, involved in the human disorder cockayne's syndrome group B. Nucleic Acids Res 1993; 21:419-426.
15. Mallery DL, Tanganelli B, Colella S et al. Molecular analysis of mutations in the CSB (ERCC6) gene in patients with Cockayne syndrome. Am J Hum Genet 1998; 62:77-85.
16. Colella S, Nardo T, Botta E et al. Identical mutations in the CSB gene associated with either Cockayne syndrome or the DeSanctis-cacchione variant of Xeroderma pigmentosum. Hum Mol Genet 2000; 9:1171-1175.
17. Horibata K, Iwamoto Y, Kuraoka I et al. Complete absence of Cockayne syndrome group B gene product gives rise to UV-sensitive syndrome but not Cockayne syndrome. Proc Natl Acad Sci USA 2004; 101:15410-15415.
18. Friedberg EC, Walker GC, Siede W. DNA repair and mutagenesis. Washington DC: American Society for Microbiology, 1995.
19. Bohr VA, Smith CA, Okumoto DS et al. DNA repair in an active gene: removal of pyrimidine dimers from the DHFR gene of CHO cells is much more efficient than in the genome overall. Cell 1985; 40:359-369.
20. Mellon I, Spivak G, Hanawalt PC. Selective removal of transcription-blocking DNA damage from the transcribed strand of the mammalian DHFR gene. Cell 1987; 51:241-249.
21. Henning KA, Li L, Iyer N et al. The Cockayne syndrome group A gene encodes a WD repeat protein that interacts with CSB protein and a subunit of RNA polymerase II TFIIH. Cell 1995; 82:555-564.
22. Nakatsu Y, Asahina H, Citterio E et al. XAB2, a novel tetratricopeptide repeat protein involved in transcription-coupled DNA repair and transcription. J Biol Chem 2000; 275:34931-34937.
23. Trzcinska AM, Girstun A, Piekielko A et al. Potential protein partners for the N-terminal domain of human topoisomerase I revealed by phage display. Mol Biol Rep 2002; 29:347-352.
24. Groisman R, Polanowska J, Kuraoka I et al. The ubiquitin ligase activity in the DDB2 and CSA complexes is differentially regulated by the COP9 signalosome in response to DNA damage. Cell 2003; 113:357-367.
25. van Gool AJ, van der Horst GT, Citterio E et al. Cockayne syndrome: defective repair of transcription? EMBO J 1997; 16:4155-4162.
26. Troelstra C, van Gool A, de Wit J et al. ERCC6, a member of a subfamily of putative helicases, is involved in cockayne's syndrome and preferential repair of active genes. Cell 1992; 71:939-953.
27. van der Horst GT, van Steeg H, Berg RJ et al. Defective transcription-coupled repair in Cockayne syndrome B mice is associated with skin cancer predisposition. Cell 1997; 89:425-435.
28. Park JS, Marr MT, Roberts JW. E. coli transcription repair coupling factor (Mfd protein) rescues arrested complexes by promoting forward translocation. Cell 2002; 109:757-767.
29. Christiansen M, Stevnsner T, Modin C et al. Functional consequences of mutations in the conserved SF2 motifs and posttranslational phosphorylation of the CSB protein. Nucleic Acids Res 2003; 31:963-973.
30. Snyderwine EG, Bohr VA. Gene- and strand-specific damage and repair in Chinese hamster ovary cells treated with 4-nitroquinoline 1-oxide. Cancer Res 1992; 52:4183-4189.
31. Tang MS, Bohr VA, Zhang XS et al. Quantification of aminofluorene adduct formation and repair in defined DNA sequences in mammalian cells using the UVRABC nuclease. J Biol Chem 1989; 264:14455-14462.
32. Wade MH, Chu EH. Effects of DNA damaging agents on cultured fibroblasts derived from patients with Cockayne syndrome. Mutat Res 1979; 59:49-60.

33. Tuo J, Muftuoglu M, Chen C et al. The Cockayne syndrome group B gene product is involved in general genome base excision repair of 8-hydroxyguanine in DNA. J Biol Chem 2001; 276:45772-45779.
34. Leadon SA, Cooper PK. Preferential repair of ionizing radiation-induced damage in the transcribed strand of an active human gene is defective in cockayne syndrome. Proc Natl Acad Sci USA 1993; 90:10499-10503.
35. Brosh RM Jr, Balajee AS, Selzer RR et al. The ATPase domain but not the acidic region of Cockayne syndrome group B gene product is essential for DNA repair. Mol Biol Cell 1999; 10:3583-3594.
36. Orren DK, Dianov GL, Bohr VA. The human CSB (ERCC6) gene corrects the transcription-coupled repair defect in the CHO cell mutant UV61. Nucleic Acids Res 1996; 24:3317-3322.
37. van Oosterwijk MF, Filon R, de Groot AJ et al. Lack of transcription-coupled repair of acetylaminofluorene DNA adducts in human fibroblasts contrasts their efficient inhibition of transcription. J Biol Chem 1998; 273:13599-13604.
38. Balajee AS, May A, Dianov GL et al. Reduced RNA polymerase II transcription in intact and permeabilized Cockayne syndrome group B-cells. Proc Natl Acad Sci USA 1997; 94:4306-4311.
39. Bradsher J, Auriol J, Proietti de Santis L et al. CSB is a component of RNA pol I transcription. Mol Cell 2002; 10:819-829.
40. D'Errico M, Parlanti E, Teson M et al. The role of CSA in the response to oxidative DNA damage in human cells. Oncogene 2007; 26(30):4336-4343.
41. Stevnsner T, Nyaga S, de Souza-Pinto NC et al. Mitochondrial repair of 8-oxoguanine is deficient in cockayne syndrome group B. Oncogene 2002; 21:8675-8682.
42. Tuo J, Jaruga P, Rodriguez H et al. The Cockayne syndrome group B gene product is involved in cellular repair of 8-hydroxyadenine in DNA. J Biol Chem 2002; 277:30832-30837.
43. Jensen A, Calvayrac G, Karahalil B et al. Mammalian 8-oxoguanine DNA glycosylase 1 incises 8-oxoadenine opposite cytosine in nuclei and mitochondria, while a different glycosylase incises 8-oxoadenine opposite guanine in nuclei. J Biol Chem 2003; 278:19541-19548.
44. Osterod M, Larsen E, Le Page F et al. A global DNA repair mechanism involving the Cockayne syndrome B (CSB) gene product can prevent the in vivo accumulation of endogenous oxidative DNA base damage. Oncogene 2002; 21:8232-8239.
45. Trapp C, Reite K, Klungland A et al. Deficiency of the Cockayne syndrome B (CSB) gene aggravates the genomic instability caused by endogenous oxidative DNA base damage in mice. Oncogene 2007; 26:4044-4048.
46. Hayashi M, Itoh M, Araki S et al. Oxidative stress and disturbed glutamate transport in hereditary nucleotide repair disorders. J Neuropathol Exp Neurol 2001; 60:350-356.
47. Ljungman M, Zhang F. Blockage of RNA polymerase as a possible trigger for uv light-induced apoptosis. Oncogene 1996; 13:823-831.
48. Ljungman M, Zhang F, Chen F et al. Inhibition of RNA polymerase II as a trigger for the p53 response. Oncogene 1999; 18:583-592.
49. Balajee AS, Proietti De Santis L, Brosh RM Jr et al. Role of the ATPase domain of the Cockayne syndrome group B protein in UV induced apoptosis. Oncogene 2000; 19:477-489.
50. Spivak G, Lloyd RS, Sweder KS. Workshop on DNA repair and related DNA transactions, a conference report. DNA Repair (Amst) 2003; 2:235-242.
51. Paunesku T, Mittal S, Protić M et al. Proliferating cell nuclear antigen (PCNA): ringmaster of the genome. Int J Radiat Biol 2001; 77:1007-1021.
52. Balajee AS, May A, Dianova I et al. Efficient PCNA complex formation is dependent upon both transcription coupled repair and genome overall repair. Mutat Res 1998; 409:135-146.
53. Balajee AS, Dianova I, Bohr VA. Oxidative damage-induced PCNA complex formation is efficient in Xeroderma pigmentosum group A but reduced in Cockayne syndrome group B-cells. Nucleic Acids Res 1999; 27:4476-4482.
54. Cleaver JE, Hefner E, Laposa RR et al. Cockayne syndrome exhibits dysregulation of p21 and other gene products that may be independent of transcription-coupled repair. Neuroscience 2007; 145:1300-1308.
55. Palitti F. Mechanisms of formation of chromosomal aberrations: insights from studies with DNA repair-deficient cells. Cytogenet Genome Res 2004; 104:95-99.
56. Proietti De Santis L, Garcia CL, Balajee AS et al. Transcription coupled repair efficiency determines the cell cycle progression and apoptosis after UV exposure in hamster cells. DNA Repair (Amst) 2002; 1:209-223.
57. Proietti De Santis L, Garcia CL, Balajee AS et al. Transcription coupled repair deficiency results in increased chromosomal aberrations and apoptotic death in the UV61 cell line, the Chinese hamster homologue of Cockayne's syndrome B. Mutat Res 2001; 485:121-132.
58. Citterio E, Van Den Boom V, Schnitzler G et al. ATP-dependent chromatin remodeling by the cockayne syndrome B DNA repair-transcription-coupling factor. Mol Cell Biol 2000; 20:7643-7653.

59. Gregory SM, Sweder KS. Deletion of the CSB homolog, RAD26, yields Spt(-) strains with proficient transcription-coupled repair. Nucleic Acids Res 2001; 29:3080-3086.
60. Fousteri M, van Hoffen A, Vargova H et al. Repair of DNA lesions in chromosomal DNA impact of chromatin structure and cockayne syndrome proteins. DNA Repair (Amst) 2005; 4:919-925.
61. Davie JR, Murphy LC. Inhibition of transcription selectively reduces the level of ubiquitinated histone H2B in chromatin. Biochem Biophys Res Commun 1994; 203:344-350.
62. Newman JC, Bailey AD, Weiner AM. Cockayne Syndrome group B protein (CSB) plays a general role in chromatin maintenance and remodeling. Proc Natl Acad Sci USA 2006; 103:9613-9618.
63. Klungland A, Hoss M, Gunz D et al. Base excision repair of oxidative DNA damage activated by XPG protein. Mol Cell 1999; 3:33-42.
64. Bradsher J, Auriol J, Proietti de Santis L et al. CSB is a component of RNA pol I transcription. Mol Cell 2002; 10:819-829.
65. Iyer N, Reagan MS, Wu KJ et al. Interactions involving the human RNA polymerase II transcription/nucleotide excision repair complex TFIIH, the nucleotide excision repair protein XPG and Cockayne Syndrome group B (CSB) protein. Biochemistry 1996; 35:2157-2167.
66. de Boer J, Hoeijmakers JH. Nucleotide excision repair and human syndromes. Carcinogenesis 2000; 21:453-460.
67. Yu A, Fan HY, Liao D et al. Activation of p53 or loss of the cockayne syndrome group B repair protein causes metaphase fragility of human U1, U2 and 5S genes. Mol Cell 2000; 5:801-810.
68. Thorslund T, von Kobbe C, Harrigan JA et al. Cooperation of the cockayne syndrome group B protein and poly(ADP-ribose) polymerase 1 in the response to oxidative stress. Mol Cell Biol 2005; 25:7625-7636.
69. Selby CP, Sancar A. Human transcription-repair coupling factor CSB/ERCC6 is a DNA-stimulated ATPase but is not a helicase and does not disrupt the ternary transcription complex of stalled RNA polymerase II. J Biol Chem 1997; 272:1885-1890.
70. Tuo J, Chen C, Zeng X et al. Functional crosstalk between hOgg1 and the helicase domain of cockayne syndrome group B protein. DNA Repair (Amst) 2002; 1:913-927.
71. Spivak G. The many faces of cockayne syndrome. Proc Natl Acad Sci USA 2004; 101:15273-15274.
72. Kraemer KH, Patronas NJ, Schiffmann R et al. Xeroderma pigmentosum, trichothiodystrophy and cockayne syndrome: a complex genotype-phenotype relationship. Neuroscience 2007; 145:1388-1396.
73. Lehmann AR. DNA repair-deficient diseases, Xeroderma pigmentosum, cockayne syndrome and trichothiodystrophy. Biochimie 2003; 85:1101-1111.
74. Jaspers NGJ, Raams A, Appeldoorn E et al. A UV-sensitive patient with a severe progeroid syndrome attributable to defective recombinational repair due to XPF/ERCC1 nuclease deficiency, workshop on DNA repair: interplay with other cellular processes. 2001; 71.
75. Bender M, Potocki L, Metry DW. Cockayne syndrome. Pediatric Dermatology 2008; 20(6)538-540.

Chapter 2

Transcription-Coupled Repair and Its Defect in Cockayne Syndrome

Kiyoji Tanaka* and Shinsuke Ito

Abstract

Cockayne syndrome (CS) is an autosomal recessive disorder clinically characterized by photosensitivity, physical and mental retardation and premature aging. It consists of two different genetic complementation groups, CS-A and CS-B. CS involves a defect in transcription-coupled nucleotide excision repair (TC-NER), a sub-pathway of nucleotide excision repair (NER) that removes transcription-blocking DNA lesions that otherwise provide a strong signal for cell death. Recent findings suggest that CSA and CSB proteins have different roles in TC-NER. The CSA protein forms an ubiquitin ligase complex consisting of DDB1, cullin4A, Roc1 and the COP9 signalosome, while the CSB protein is a member of the SWI2/SNF2 family and has DNA-dependent ATPase and ATP-dependent chromatin remodeling activities. It has been shown that xeroderma pigmentosum (XP) patients with CS (XP/CS) have defects in transcription as well as in NER. It is suggested that CS is a disorder with DNA repair/transcription-deficiencies.

Introduction

Nucleotide excision repair (NER) is a multiprotein DNA repair system that can remove a wide variety of DNA helix-distorting lesions, including UV-induced damage such as cyclobutane pyrimidine dimers (CPD) and 6-4 photoproducts (6-4PP), intra-strand crosslinks and protein-DNA crosslinks.[1] NER is well conserved from *Escherichia coli* to mammals and consists of the consecutive steps of damage recognition, dual incisions on either side of the damage, excision of 24-32 oligonucleotides containing the damage, gap filling by repair DNA synthesis using the error-free strand as a template and ligation (Fig. 1).

The biological importance of NER in humans has been suggested by the presence of autosomal recessive genetic disorders with a defect in NER, such as xeroderma pigmentosum (XP), Cockayne syndrome (CS), cerebro-oculo-facio-skeletal syndrome (COFSS), UV-sensitive syndrome ($UV^{S}S$) and trichothiodystrophy (TTD).[2] XP patients are hypersensitive to sunlight and show an increased incidence of skin cancers in sunlight-exposed areas. Some XP patients show neurological and physiological complications. CS patients are sensitive to sunlight, but have no predisposition to sunlight-induced skin cancer; instead, they show severe developmental and neurological abnormalities as well as premature aging. COFSS is regarded as a more severe form of CS. TTD patients show sulfur-deficient brittle hair in addition to CS features. $UV^{S}S$ patients show only mild photosensitivity and freckles without skin cancers or any neurological or physiological abnormalities. Seven NER-deficient complementation groups have been identified in XP (XP-A

*Corresponding Author: Kiyoji Tanaka—Human Cell Biology Group, Graduate School of Frontier Biosciences, Osaka, University and Solution-Oriented Research for Science and Technology (SORST), Japan Science and Technology Agency (JST), 1-3 Yamadaoka, Suita, Osaka, 565-0871, Japan. Email: ktanaka@fbs.osaka-u.ac.jp

Molecular Mechanisms of Cockayne Syndrome, edited by Shamim I. Ahmad.

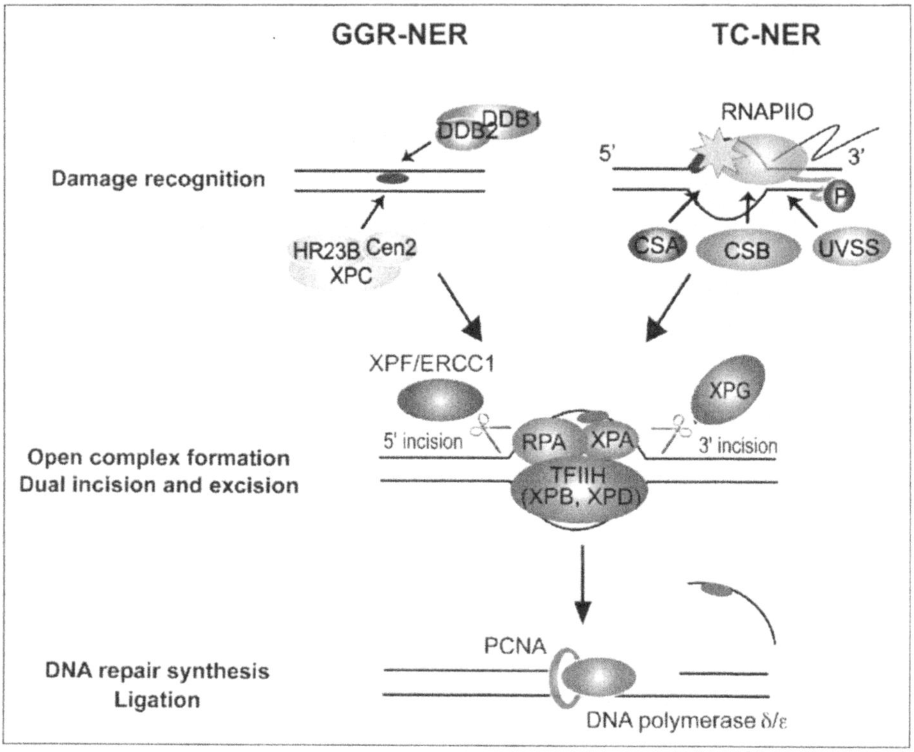

Figure 1. Two subpathways in nucleotide excision repair. The major difference between GG-NER and TC-NER is derived from their damage recognition step (described in the text). In GG-NER, UV-DDB could play a role in recruiting the XPC complex. In TC-NER, blockage of the elongating RNA polymerase IIO at the site of DNA damage serves as the damage recognition. CSA, CSB and UV^{S}S might be required for the TC-NER-specific pathway. After damage recognition by a specific pathway, common factors are adopted to excise damaged nucleotides. The resulting gap is filled by DNA polymerase and finally rejoined by DNA ligase to complete the reaction.

to XP-G) and two in CS (CS-A and CS-B). XP variant (XP-V) cells are proficient in NER but defective in translesion DNA synthesis.[3] Moreover, certain patients with mutations in *XPB*, *XPD*, *XPF*, *XPG* or *ERCC1* show features of CS in addition to symptoms of XP (XP-B/CS, XP-D/CS, XP-G/CS, XFE, ERCC1/COFSS). Two UV^{S}S patients have a homozygous null mutation in *CSB*, while the others had mutation(s) in an unidentified gene.[4] About half of the TTD patients exhibit photosensitivity and NER-deficiency. The photosensitivity of TTD has been associated with mutations in *XPB*, *XPD* or *TTDA*, which encode three of the ten subunits that constitute the DNA repair/transcription factor TFIIH.[5] *TTDN1* was identified as the gene responsible for the nonphotosensitive form of TTD.[6]

NER operates via two pathways: global genome nucleotide excision repair (GG-NER) and transcription-coupled nucleotide excision repair (TC-NER).[1] (Fig. 1). GG-NER can act on DNA lesions at any location in the genome, while TC-NER is involved in rapid removal of DNA lesions on the transcribed strand in transcription-active regions followed by a resumption of transcription. It is known that CS-A, CS-B and UV^{S}S cells are specifically deficient in TC-NER but proficient in GG-NER, while XP-C and XP-E cells are deficient in GG-NER but proficient in TC-NER.

In the GG-NER pathway, the XPC/HR23B/Cen2 complex and UV-DDB recognize the helical distortion caused by DNA damage.[7,8] It has been suggested that translocating RNA polymerase IIO, which has a hyperphosphorylated CTD in the largest subunit (Rpb1) of the polymerase complex, stalls at DNA lesions and acts as a the damage recognition factor in TC-NER. Following each DNA damage-recognition step, TC-NER and GG-NER converge into a common pathway.[14] The core NER factors, including TFIIH, XPG, XPA and RPA, are recruited to the damaged sites, leading to the local unwinding of the DNA double helix (open complex formation), verification of actual damage and identification of the damaged strand.[9,10] DNA-dependent ATPase and helicase activities of the XPB and XPD subunits of TFIIH play a crucial role in the formation of the open complex.[11] Two structure-specific endonucleases, XPF/ERCC1 and XPG, subsequently introduce single-stranded breaks on the 5′ and 3′ sides of the open complex respectively, leading to excision of the oligonucleotides containing the damage. Then, DNA polymerase δ (delta) and DNA ligase III/XRCC1 are employed in gap filling and religation, while in proliferating cells, DNA polymerase ε (epsilon) and DNA ligase I are also used in these steps.[12] Recently, DNA polymerase κ (kappa) was suggested to be involved in NER as well.[13]

In this chapter, we discuss findings on the molecular mechanisms of defective TC-NER in CS. It has been proposed that the CS phenotype may not always be directly linked to defects in TC-NER. CS factors such as CSA, CSB, TFIIH and XPG are involved in transcription as well as TC-NER, hence a defect in transcription per se can be linked to the development of CS. The causative relationship between the CS phenotype and defects in transcription, as well as in TC-NER, is also discussed.

Defective TC-NER in Cockayne Syndrome

TC-NER in Mammalian Cells

Hanawalt et al have discovered that certain silent regions of the mammalian genome are refractory to repair, while transcribed genes are subjected to a dedicated pathway in which the transcribing DNA strands are preferentially repaired.[15] This pathway is termed transcription-coupled repair (TCR) or transcription-coupled nucleotide excision repair (TC-NER). It is known that certain types of DNA lesions on the transcribed strand, such as CPD, interfere with transcription and that the persistent blockage of transcription by RNA polymerase IIO leads to reduced transcription of critical genes and acts as a potential signal for apoptosis and induction of the p53 response.[16] To prevent apoptosis and collapse of replication forks by stalled RNA polymerase IIO, TC-NER removes rapidly the transcription-blocking lesions. The findings that CS-A, CS-B and UV^{S}S cells are deficient in TC-NER but proficient in GG-NER indicate that *CSA, CSB* and *UV^{S}S* are selectively required for TC-NER.[17,18] The resumption of RNA synthesis after exposure to UV-irradiation and certain chemical agents that induce bulky DNA adducts is absent in CS-A, CS-B and UV^{S}S cells. These results suggested that the resumption of RNA synthesis after UV-irradiation is dependent on *CSA, CSB* and *UV^{S}S.*

The CSA gene encodes a 44-kDa protein containing WD-40 repeat motifs that are involved in protein-protein interactions and protein complex formation.[19] The CSB protein is a member of the SWI2/SNF2 family and has DNA-dependent ATPase and ATP-dependent chromatin remodeling activities.[20] In *E. coli*, the Mfd protein is a transcription-repair coupling factor (TRCF), that also belongs to the SWI2/SNF2 superfamily and has DNA-dependent ATPase activity.[21] Mfd is sufficient for TC-NER in *E. coli*; it releases RNA polymerase stalled at DNA lesions and recruits the repair factor UvrABC endonuclease at damaged sites. Although CSB is a member of the SWI2/SNF2 family and has DNA-dependent ATPase activity, it does not remove the stalled RNA polymerase II elongation complex from the DNA template; its role is to recruit other TC-NER factors.[22,23]

UV-Induced Translocation of CSA to the Nuclear Matrix

It has been reported that CSA is translocated to the nuclear matrix and it colocalizes with RNA polymerase IIO in UV-irradiated cells.[24] Furthermore, the translocation is dependent on a functional CSB, TFIIH, chromatin structure and transcription elongation.[25] The nuclear matrix is assumed to play an important role in nuclear metabolism. RNA polymerase IIO associates with the nuclear matrix;[26] UV-induced repair patches are enriched in the nuclear matrix and this phenomenon was enhanced in TC-NER-proficient XP-C cells and abolished in TC-NER-deficient CS-B-cells.[27,28] These results suggested that TC-NER occurs in the nuclear matrix and that the UV-induced translocation of CSA to the nuclear matrix is relevant to TC-NER.

Assembly of the TC-NER Complex at a Chromatin-Bound RNA Polymerase IIO That Is Stalled at UV-Damaged Sites

As described above, blockage of transcription by RNA polymerase IIO at the site of DNA damage is thought to trigger TC-NER. In order to elucidate TC-NER mechanisms, transcription elongation complexes, including RNA polymerase IIO stalled at UV-damaged sites, were isolated using in vivo crosslinking and chromatin immunoprecipitation (ChIP).[23] UV-dependent association of CSA and CSB with the stalled RNA polymerase IIO was detected. Also it was found that CSA and CSB play distinguishable roles in the formation of the TC-NER complex (Fig. 2). The interaction between CSB and RNA polymerase IIO was enhanced in UV-irradiated normal cells, suggesting CSB is recruited to the chromatin-bound RNA polymerase IIO that is stalled at UV-damaged sites. CSA was associated with the chromatin-bound RNA polymerase IIO in a CSB-dependent manner, consistent with the results mentioned above.[24,25]

Another TC-NER factor, XAB2 (XPA-binding protein 2), was isolated by virtue of its ability to interact with XPA. It contains a tetratricopeptide repeats (TPR) motif responsible for protein complex formation and/or protein interactions and it interacts with CSA, CSB and RNA polymerase II as well as with XPA.[29] In addition, XAB2 forms a stable protein complex consisting of hAquarius, XAB2, hPRP19, CCDC16, hISY1 and PPIE, which is involved in pre mRNA splicing and transcription as well as in TC-NER.[30] Experiments with ChIP revealed that XAB2 was recruited to the chromatin-bound RNA polymerase IIO stalled at UV-damaged sites. This recruitment is dependent on CSA and CSB (Fig. 2).[23] XAB2 could be a factor for tethering transcription, TC-NER and pre mRNA splicing.

Core NER factors such as XPA, RPA, ERCC1, TFIIH and XPG and the chromatin remodeling factor p300 were also recruited to the chromatin-bound RNA polymerase IIO; this was dependent on CSB but not on CSA (Fig. 2). The CSA complex may not be required for the recruitment of the proper NER factors to the stalled RNA polymerase IIO. The nucleosomal binding protein HMGN1 was also recruited to the chromatin-bound RNA polymerase IIO, requiring CSA and CSB. Although chromatin is opened in the transcription-active region, further modification of the chromatin structure by chromatin remodeling factors may be required for TC-NER. The transcription factor TFIIS has been implicated in TC-NER and the resumption of transcription, via the cleavage of protruding mRNA to generate a new 3′ end at the catalytic center of RNA polymerase IIO.[31,32] The interaction of TFIIS with the stalled RNA polymerase IIO is dependent on CSA and CSB. On the other hand, the GG-NER-specific factor XPC has not been detected in the stalled RNA polymerase IIO complex, indicating that the functional TCR complexes assembled in the intact cell nucleus were specifically isolated and that TC-NER occurs without disruption of the stalled RNA polymerase IIO/template DNA/nascent mRNA ternary complex. Taken together, CSB may be the first factor to be recruited to the stalled RNA polymerase IIO at damaged sites, acting as a transcription-repair coupling factor.

DNA Damage-Induced Ubiquitination of RNA Polymerase IIO and CSB

CSA or DDB2 can form a multiprotein complex that contains DDB1, cullin4A and Roc1, that displays ubiquitin ligase activity in vitro.[33] Both the CSA and the DDB2 complexes contain the COP9 signalosome (CSN), which possesses ubiquitin isopeptidase activity and inhibits the ubiquitin ligase activity of these two complexes in vitro. However, with regard to their association with CSN, the CSA and DDB2 complexes responded to UV-induced DNA damage differently. CSN disappeared from the DDB2 complex shortly after UV-irradiation and NEDD8 was conjugated on cullin4A, presumably leading to the activation of ubiquitin ligase activity in the DDB2 complex. CSN reassociated with the DDB2 complex 2 h after UV-irradiation and NEDD8 was

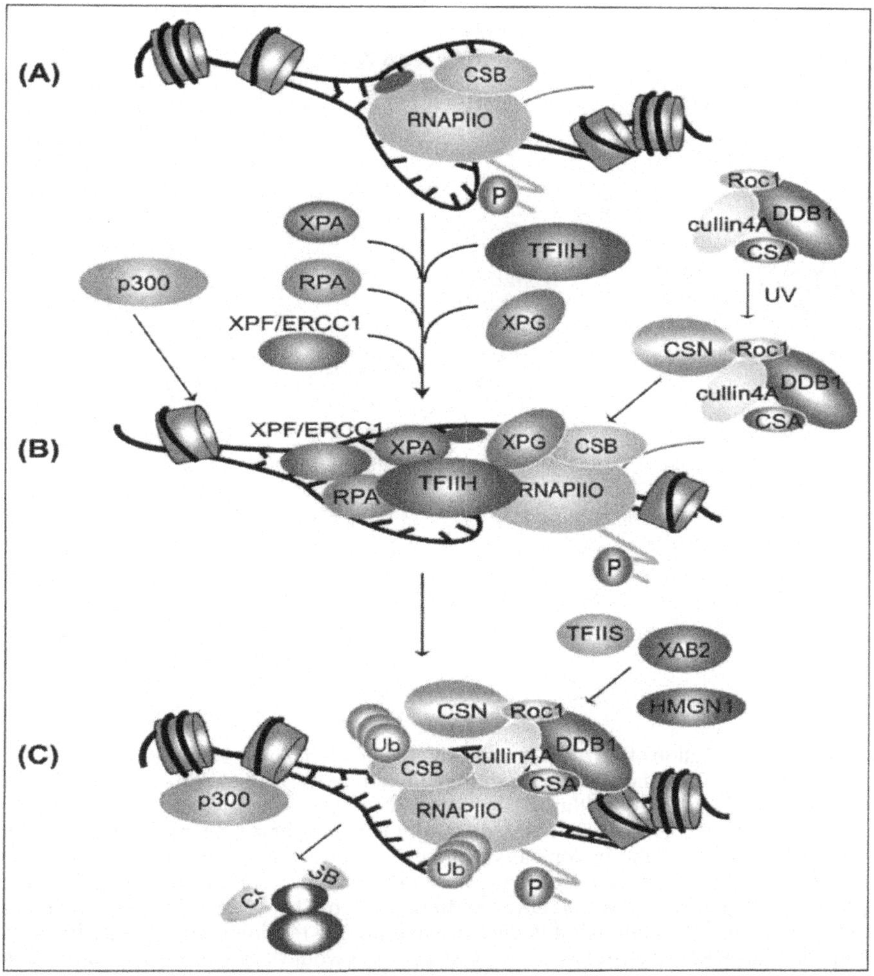

Figure 2. Sequential assembly of the TC-NER complex. A) RNA polymerase IIO encounters a lesion on the transcribed strand and transcription is stalled. B) CSB might be engaged in chromatin remodeling and modulate the interface between RNA polymerase IIO and DNA in concert with XPG and TFIIH, which could allow the lesion to be accessed and then excised by core NER factors. In addition, HAT p300 and the CSA ubiquitin ligase complex are recruited in a CSB-dependent manner. Following UV damage, the CSA complex associates with the COP9 signalosome (CSN), which would render ubiquitin ligase an inactive form. C) CSA would be required not only for the assembly of TFIIS, HMGN1 and XAB2 but also for ubiquitination of both RNA polymerase IIO and CSB, resulting in the resumption of transcription after removal of the damaged DNA.

deconjugated from cullin4A. In contrast, the amount of CSN associated with the CSA complex increased 30 min after UV-irradiation and then declined. Regardless of UV-irradiation, no conjugates of NEDD8 with cullin4A were detected in the CSA complex.[33] These results suggest that the ubiquitin ligase activity in the CSA complex is rapidly inactivated after UV-irradiation by the recruitment of CSN and then restored at a later time. However, the siRNA-based knockdown of CSN5, which is one of the subunits of CSN with isopeptidase activity and deconjugates NEDD8

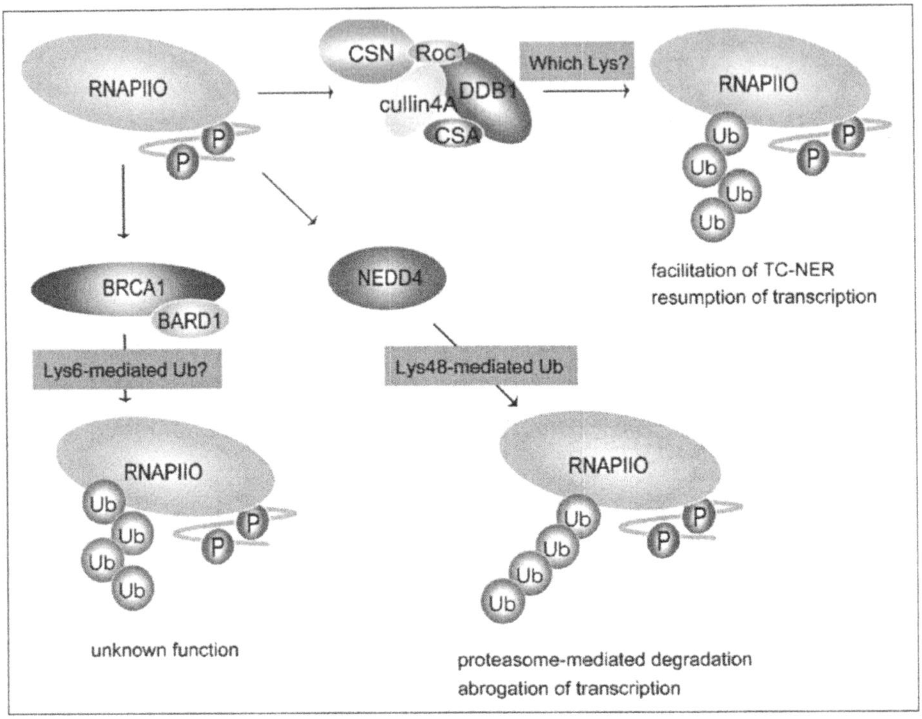

Figure 3. Ubiquitination of RNA polymerase IIO is mediated by at least three different ubiquitin ligases following DNA damage. At least three ubiqutin ligases ubiquitinate RNA polymerase IIO following DNA damage, although the fate of RNA polymerase IIO and the outcome might be different. The BRCA1/BARD1 heterodimer ubiquitinates RNA polymerase IIO via a lysine (Lys) 6-linked manner; the biological significance of this process remains unclear. NEDD4 ubiquitinates RNA polymerase IIO in a Lys48-linked manner, thereby promoting degradation by the proteasome, which would abrogate transcription. The CSA E3 complex ubiquitinates RNA polymerase IIO, although it is unclear which lysine residue is engaged in the process. Ubiquitination of RNA polymerase IIO by CSA complex would facilitate the reaction of TC-NER and resumption of transcription.

from cullin4A in vitro, led to defects both in TC-NER and in GG-NER.[33] Taken together, these results suggested a more intricate in vivo regulation of the ubiquitin ligase activity of the CSA and DDB2 complexes by CSN in TC-NER and GG-NER.

It has been shown that the largest subunit (Rpb1) of RNA polymerase IIO is polyubiquitinated shortly after UV-irradiation in normal human cells, but not in CS-A and CS-B-cells, indicating that the UV-induced polyubiquitination of RNA polymerase IIO is dependent on CSA and CSB.[34] Consistent with these findings, RNA polymerase IIA (hypophosphorylated form) was detected in the CSA complex in a soluble chromatin fraction purified from nonirradiated cells, while RNA polymerase IIO was associated with the CSA complex after UV-irradiation and dissociated 4 h after the irradiation.[33] These results suggest that RNA polymerase IIO, stalled at UV-damaged sites, is polyubiquitinated by the CSA complex.

The BRCA1/BARD1 heterodimer is also an ubiquitin ligase that ubiquitinates RNA polymerase IIO and the overexpression of BRCA1 in cells stimulated the UV-induced ubiquitination of RNA polymerase IIO[35] (Fig. 3). However, it was recently reported that neither the CSA complex nor BRCA1/BARD1 ubiquitin ligase are directly involved in the ubiquitination of RNA polymerase IIO in response to UV damage.[36] Instead, Nedd4, the human homolog of budding yeast ubiquitin ligase Rsp5, is required for the ubiquitination and degradation of RNA polymerase IIO in response to blocking of transcription elongation caused by UV-damage in vivo. It was indicated that the effect of CS-deficiency on ubiquitination of RNA polymerase IIO is indirect due to the strong decrease in general transcription after UV-damage in CS-A and CS-B-cells. Likewise, UV-induced ubiquitination of RNA polymerase IIO was reduced when transcription was inhibited by the depletion of TFIIB. However, the studies were carried out with fractions containing lysine 48-linked polyubiquitinated RNA polymerase IIO purified by its binding to an UBA domain (GST-Dsk2 resin).[36] It is known that the BRCA1/BARD1 heterodimer is involved in lysine 6-linked ubiquitination.[37] It is, however, not yet known which lysine residue of ubiquitin is involved in the CSA complex-dependent ubiquitination. Therefore, different ubiquitin ligases might be involved in different lysine-linked ubiquitinations of RNA polymerase IIO for different biological purposes (Fig. 3). NEDD4-knockdown cells are not sensitive to UV-irradiation, indicating that the lysine 48-linked polyubiquitination of RNA polymerase IIO by NEDD4 is not directly involved in TC-NER. It may be involved in ubiquitin-mediated degradation of the arrested RNA polymerase IIO by the proteasome when TC-NER has failed.

It was recently reported that CSB physically interacts with and is polyubiquitinated by the CSA complex.[38] The polyubiquitinated CSB is degraded by the proteasome 3-4 h after UV-irradiation (Fig. 2). It was suggested that CSA-dependent polyubiquitination and degradation of CSB are required for the recruitment of the core NER factors to the RNA polymerase IIO-stalled sites in TC-NER and/or for resumption of transcription after UV-irradiation.

Reconstitution of TC-NER Reactions In Vitro

Although the stalled RNA polymerase IIO is thought to provide an efficient signal for damage recognition in TC-NER, it may prevent core NER factors from accessing the lesion and need to be removed in order for DNA repair to occur. In fact, the RNA polymerase IIO that is stalled at a CPD completely covers the damaged-site and prevents access to a photolyase that catalyzes monomerization of the damage.[39] However, other in vitro studies with reconstituted incision systems, using recombinant NER factors or nuclear extracts, indicated that dual incision by XPG and XPF/ERCC1 endonucleases occurred even in the presence of RNA polymerase IIO stalled at a cisplatin lesion during transcription.[40] RNA polymerase IIO might remain associated with the incised damaged DNA template. However, the stalled RNA polymerase IIO was released by incubation with HeLa cell extracts along with ATP, but the release was independent of dual incision and of the functions of ATP-dependent CSB and TFIIH.

It has been reported that CSB binds to RNA polymerase II and recruits TFIIH. Recently, it has been shown that XPG associates with RNA polymerase IIO stalled by nucleotide deprivation or cisplatin lesions and binds to a DNA bubble that resembles bubbles associated with transcription, 14-22 nt long.[41] The R domain (or the spacer region) and C-terminal domain of XPG are responsible for binding to the transcription-sized DNA bubble, but neither domain is directly involved in the incision reaction. XPG and CSB form a stable complex with the transcription-sized DNA bubble. The ATPase activity of CSB is enhanced in the presence of bubble DNA and the C-terminal domain of XPG enhances the bubble DNA-dependent ATPase activity of CSB. These results suggest that XPG and CSB act cooperatively on the transcription bubble to which RNA polymerase IIO is bound during TC-NER. On the other hand, XPG incises a 15-nt DNA bubble, which resembles the transcription bubble. However, the incision was blocked by prior addition of RNA polymerase IIO in the experimental system and the inhibition was released by the addition of TFIIH and ATP, although the restoration of incision was not accompanied by the release of bound RNA polymerase II. These results suggest that XPG is involved in the recognition of stalled RNA polymerase IIO and the stimulation of CSB activity, leading to the recruitment of TFIIH at the site of the stalled RNA polymerase IIO, which is then remodeled via an ATP hydrolysis-dependent process to allow XPG to incise the bubble junction.

The reconstitution of an in vitro TC-NER system was attempted using the damaged DNA template with stalled RNA polymerase IIO.[14] The damaged template contained both a promoter and a cisplatin DNA adduct. When the DNA template was incubated with a reconstituted transcription system containing RNA polymerase II and basal transcription factors TBP, TFIIB, TFIIE, TFIIF and TFIIH, a truncated transcript, stalled at the cisplatin lesion, was produced, while a full-length transcript was produced by undamaged DNA template. On the other hand, 26-34 nt dual incision products were released by the incubation of the damaged DNA template with a reconstituted incision system containing XPC/HR23B, TFIIH, XPA/RPA, XPG and XPF/ERCC1.[14] Thus, the DNA template was competent for both transcription and incision and was used in the reconstituted TC-NER system as shown below.[14]

To isolate the RNA polymerase IIO stalled at the damaged DNA template, a transcription reaction was performed using the damaged DNA template in the reconstituted transcription system and the transcription reaction sample was washed with buffer containing appropriate concentrations of salt and detergent.[14] The purified stalled RNA polymerase IIO was then incubated with whole cell extracts from GG-NER deficient XP-C cells. Most of the NER factors, including XPG, TFIIH, XPA, RPA and XPF/ERCC1 were recruited to the stalled RNA polymerase IIO in the presence of ATP. Incubation of the stalled RNA polymerase IIO in a reconstituted incision system without XPC did not release incision products. However, when CSB was added in a reconstituted incision system without XPC, small but significant amounts of released incision products were detected in the presence of ATP. Thus, the RNA polymerase IIO-mediated incision reaction is CSB-and NER factor-dependent but XPC-independent. However, the yield of the incision reaction was only 1-2%, suggesting that several other factors such as the CSA complex, XAB2 and MMS19, are missing in the in vitro TC-NER system. The role of ATP hydrolysis by CSB could be to remodel the interface between the DNA template and the NER factors, leading to stabilization of RNA polymerase IIO for the recruitment of NER factors. Dephosphorylation of RNA polymerase IIO, or mutant CSB derived from CS-B patients, resulted in a decrease in incision, suggesting that the in vitro TC-NER reaction relied on RNA polymerase IIO and CSB. It was shown that, similar to what was observed in GG-NER, TFIIH and XPA were first recruited to the RNA polymerase IIO, while XPG and XPF were recruited later.[42] The addition and recruitment of NER factors led to the increased release of RNA polymerase IIO from the DNA template. The release was dependent at least partially on the helicase activity of TFIIH in the presence of ATP, but not on CSB in this experimental condition.[14]

Defective Repair of Topoisomerase I DNA Covalent Complexes in Cockayne Syndrome

It has been reported that CS-A and CS-B-cells are hypersensitive to killing by a DNA topoisomerase I-inhibitor, camptothecin (CPT) and that recovery of DNA synthesis and RNA synthesis after CPT-treatment is decreased in CS-A and CS-B-cells.[43] DNA topoisomerase I (Top1) reversibly nicks DNA and relaxes superhelical tension in DNA induced during replication and transcription and generates a reversible and transient intermediate known as the Top1~DNA covalent complex (Top1-cc), in which Top1 is covalently attached via a tyrosyl residue (human Tyr723) to the 3'-terminus of a single-stranded nick.[44] Following release of torsional stress, Top1 reseals the nick and restores the integrity of the double helix religation. However, CPT is known to block the religation reaction of Top1, resulting in accumulation of Top1-cc. In addition, collision of Top1-cc with replication forks results in abortive Top1-associated DNA double stranded breaks, while collision of Top1-cc with the transcriptional machinery converts Top1-cc into abortive Top1-associated DNA single stranded breaks, which can block transcription. These results indicated that there are CSA and CSB-dependent pathways to repair a Top1-associated DNA lesion and resume transcription after CPT treatment. Although the exact roles of CSA and CSB in the repair of Top1-associated DNA lesions and resumption of transcription are not known, the CS features may be caused at least partially by a failure in the repair of these lesions and resumption of transcription.

Defective Transcription in Cockayne Syndrome

CSB Protein Functions in Transcription

Recent studies provide evidence that CSB has an impact on transcription. In addition to its DNA-dependent ATPase and chromatin remodeling activities, CSB locally influences the DNA conformation, most likely by wrapping the DNA around itself,[46] thereby modulating the interface between RNA polymerase II and DNA. CSB stimulates transcriptional elongation by both RNA polymerase I and RNA polymerase II in vitro.[47,48] CSB is included in a protein complex containing RNA polymerase I and histone methyltransferase G9a and is recruited to both the promoter and pre rRNA coding regions, thereby mediating active transcription by RNA polymerase I.[49] In addition, CSB interacts with transcriptional complexes including TFIIH, TFIIE, p53, RNA polymerase II and RNA polymerase I as well as NER factors such as XPG, XPA and XAB2.[22,29,48,50,51]

A transcriptome analysis using expression arrays and a comparative expression analysis revealed that the expression of wild-type CSB cDNA in CS-B-cells induced significant changes in gene expression.[52] Many of the genes regulated by CSB were also affected by inhibitors of histone deacetylase or poly (ADP-ribose) polymerase, which affect chromatin structure. Therefore, CSB has a role in transcription that is at least partly through chromatin maintenance and remodeling via its SWI2/SNF2 ATPase activity.

Another line of evidence implicates CSB in the formation of the transcription initiation complex after cells are damaged with UV.[53] One of the hallmarks of CS cells is a defect in the recovery of RNA synthesis after UV-irradiation. It has been postulated that this transcriptional defect is derived from the cellular inability of TC-NER to repair the UV damage on the transcribed strand. However, this is not the case because CS-B-cells cannot transcribe un-damaged exogenous genes if cells are previously UV-irradiated. Moreover, the study using chromatin immunoprecipitation (ChIP) revealed that the occupancy of not only RNA polymerase II but also general transcription factors on the promoter of housekeeping genes was reduced in CS-B-cells after UV-irradiation, leading to inhibition of transcription initiation. Consistent with such observations, earlier reports have demonstrated that extracts prepared from UV-irradiated CS-B as well as CS-A cells exhibited reduced transcriptional activity at all times after UV-irradiation, while normal cells showed fully restored activity after a transient drop.[54] The reduction of transcriptional activity in extracts of UV-irradiated CS cells coincided with persistent depletion of hypophosphorylated RNA polymerase II (RNA polymerase IIA), which participates in the initiation of transcription. Taken together, CSA and CSB might recycle the phosphorylation status of RNA polymerase II via unknown mechanisms to resume transcription after UV-irradiation.

In contrast to the obvious roles of CSB in transcription, the molecular functions of CSA in transcription remain to be elucidated. However, as described above, CSA forms a complex with DDB1, cullin4A and Roc1 and displays ubiquitin ligase activity.[33] It has become evident that the ubiquitin-proteasome pathway is involved in nuclear receptor-mediated transcription via proteolysis, facilitating the turnover of transcription complexes, including general transcription factors and nuclear receptors.[55] The turnover of these factors promotes cyclic transcription. As described earlier, the CSA ubiquitin ligase complex degrades CSB following UV-irradiation in a proteasome-dependent manner, allowing RNA polymerase II to resume transcription.[38] These results suggest that CSA helps to remove the factors that block RNA polymerase II so that transcription can be resumed. Further studies will shed light on the role of CSA in transcription.

Defective Transcription and Transctivation in XP-D and XP-D/TTD Patients

The exact molecular pathogenesis of CS features remains to be elucidated and may not be exclusively linked to a defect in TC-NER. This idea is supported by the notion that (i) mutations of the *XPA* gene, which is indispensable for both TC-NER and GG-NER, do not cause CS and (ii) although $Xpa^{-/-}$ and $Xpg^{-/-}$ mice shared defects in both TC-NER and GG-NER, only $Xpg^{-/-}$ mice showed CS-like features such as postnatal growth failure, neurological abnormalities and early death before weaning.[56] CS may be associated with a defect in a process other than TC-NER. One clue to understanding the pathogenesis of the CS features comes from the evidence that in rare cases, mutations in *XPB*, *XPD* and *XPG* cause features of CS combined with XP (XP-B/CS, XP-D/CS and XP-G/CS).

XPB and XPD are subunits of TFIIH, a multifunctional complex involved in basal transcription, transactivation, the cell cycle and NER.[57] The complexity of clinical phenotypes among three disorders (XP, XP/CS and TTD) associated with defects in TFIIH presumably reflects the magnitude of molecular defects of TFIIH in multiple functions.[58] TFIIH can be divided into two subcomplexes; the core TFIIH (IIH6) composed of XPB, p62, p52, p44, p34 and p8 and a cdk-activating kinase (CAK) subcomplex that contains cdk7, cyclin H and MAT1. The two subcomplexes are bridged by XPD.[59,60] In vitro reconstitution of TFIIH, using recombinant subunits, suggests that not all the enzymatic activities and/or subunits are constantly required for a given function of the TFIIH complex.[61] In NER, XPB and XPD helicases co-operate in the unwinding of duplex DNA around lesions to allow the recruitment of NER factors to the damaged site. In basal transcription, TFIIH functions in concert with at least four other basal transcription factors. In addition to having an indispensable role in NER, XPB helicase is essential for transcription initiation and promoter escape,[62] steps regulated by phosphorylation in CTD of the largest subunit of RNA Polymerase II mediated by the cdk7 kinase of TFIIH.[63] In contrast, XPD helicase is dispensable for transcription although XPD facilitates optimal transcription by anchoring the CAK subcomplex to the core TFIIH.[64,65] Electron microscopic and biochemical analyses suggest that TFIIH formed a ring-like architecture with the bulge composed of CAK, whereas XPD, which directly interacts with p44 and MAT1, might be located in the interface between the core TFIIH and CAK.[66,67] The functional significance of XPD is manifested by the fact that some mutations found in XP-D patients prevent the interaction of XPD with p44, resulting in the dissociation of CAK and/or XPD from the core TFIIH.[68]

The effects of *XPD* mutations derived from XP-D and XP-D/TTD patients on in vitro NER and basal transcription systems were examined.[65] The recombinant TFIIH, in which the XPD subunit has amino acid changes derived from XP-D patients, are deficient in XPD helicase activity and all the XP-D and XP-D/TTD patients' cells were deficient in dual incision activity. On the other hand, TFIIH containing *XPD* mutations derived from XP-D/TTD patients exhibited a defect in in vitro basal transcription in addition to a reduced intracellular concentration of TFIIH. Moreover, it has been reported that CAK, within TFIIH, phosphorylates numerous transcriptional activators including nuclear receptors (NRs) such as retinoic acid receptor (RAR α and γ), estrogen receptor (ER α), peroxisome proliferator-activated receptor (PPAR α and γ), and regulator of vitamin D receptor (Ets1) and that the ligand-dependent phosphorylation of NRs was absent and thus the transactivation was not induced in XP-D and XP-D/TTD[69-73] (Fig. 4). Indeed, it has been shown that the hypoplasia of adipose tissues in XP-D/TTD mice is due to a defective transactivation of nuclear receptors, like PPARs, which are essential for ligand-dependent control of the activation of hormone-responsive genes, thus allowing lipid metabolism, differentiation and the survival of adipocytes in vivo.[72] Given that CS and TTD patients show similar phenotypes such as postnatal growth failure, neurological abnormalities, loss of subcutaneous fat tissue, hypogonadism, the CS features may be caused by an abnormality in the transcriptional process.

Defective Transactivation in XP-G/CS

XPG encodes a structure-specific endonuclease required for making the 3′ incision during NER. As discussed before, in TC-NER, XPG is involved in the recognition of the stalled RNA polymerase IIO, stimulation of CSB activity and recruitment of TFIIH to the site of the stalled RNA polymerase IIO, which is then remodeled in an ATP hydrolysis-dependent process to allow XPG to incise the 3′ junction of a transcription sized bubble.

The question therefore remains as to why some mutations in the *XPG* gene manifest the features of CS in XP-G/CS patients.[74] The *XPG* mutations found in XP-G/CS patients are expected to produce XPG with a C-terminal deletion, whereas the mutations in patients with mild XP-G produce a full-length XPG that is somewhat affected in its endonuclease activities.[75,76] indicating that the deletion of the C-terminal region of XPG is responsible for the phenotype of CS. Reports on *Xpg* mice support such causal relationship.[77,78] Mice with the *Xpg*-null mutation (*Xpg*$^{-/-}$) or mice missing the C-terminal 360 amino acids of XPG (*Xpg*D811stop) exhibited postnatal growth retardation, marked degeneration of Purkinje cells, apoptotic cells in the granule cell layer of the cerebellum and a shortened life span, whereas mice with a missense mutation in a highly conserved amino acid (*Xpg*D811A), which completely abolishes the 3′ incision activity, exhibited normal growth and a normal life span.[77,78] Since *Xpa*$^{-/-}$ mice, which are completely defective in both TC-NER and GG-NER, do not show such physical and neurological abnormalities, these observations strongly suggest that XPG has an additional function other than NER, which might be strongly associated with growth and development. In line with the suspected connection of CS with transcription, Rad2, the yeast counterpart of XPG, has a role in RNA polymerase II transcription, although the precise functions of Rad2 and the human XPG protein have not yet been elucidated.[79] A recent study demonstrates that XPG forms a stable protein complex with TFIIH and stabilizes TFIIH.[80] Interestingly, truncation mutations in *XPG*, found in XP-G/CS patients, that prevent the association with TFIIH, also resulted in the dissociation of CAK and XPD from the core TFIIH, whereas point mutations in the active site of the endonuclease, giving rise to pure XP, do not affect the ability to form a complex, thereby stabilizing TFIIH (Fig. 4). As a consequence, the transactivation of nuclear

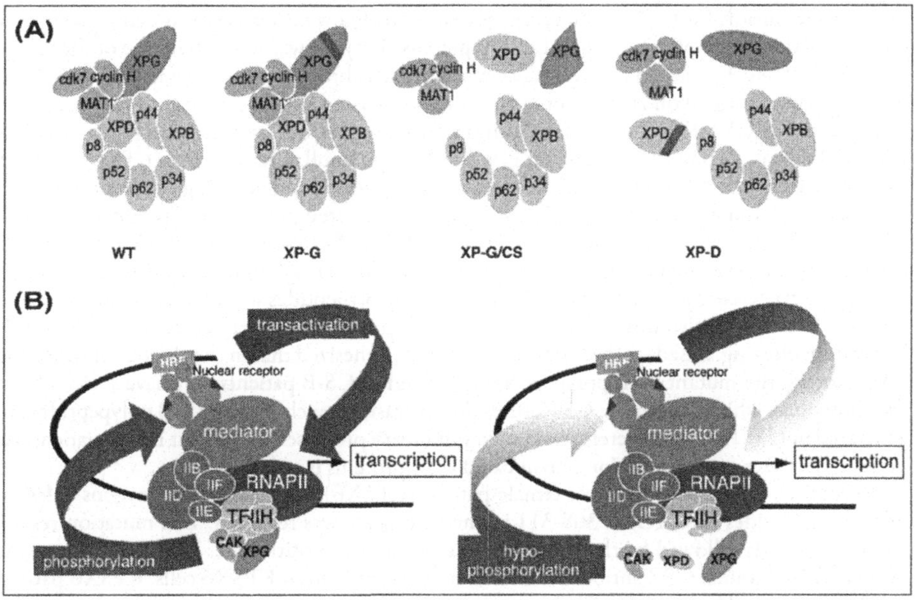

Figure 4. Schematic model for the role of XPG and XPD in maintaining the integrity of TFIIH and transactivation. A) XPG forms a stable complex with TFIIH and functions in maintaining the integrity of TFIIH. Full-length XPG with a missense mutation in the nuclease domain that was found in a patient with mild XP-G forms a complex with TFIIH similar to that formed by WT XPG, while a mutant XPG with a C-terminal deletion derived from a XP-G/CS patient causes the dissociation of CAK and XPD from the core TFIIH. Mutations in XPD that disturb the interaction with p44 and/or XPG lead to the dissociation of the CAK subunit. B) The XPG-TFIIH complex is involved in phosphorylation of hormonal receptors, such as the estrogen receptor, which is a crucial step for the ligand-dependent transactivation (left). Some mutations in either XPD or XPG destabilize the interface between the core TFIIH and CAK and thus decrease cdk7 kinase activity towards its nuclear receptor during transactivation. As a consequence, some XP-D and XP-G patient cells have a defect in this process due to the formation of an unstable TFIIH architecture (right).

receptors was disturbed in XP-G/CS cells as well as in $xpg^{-/-}$ mouse embryonic fibroblasts, due to hypophosphorylation of NRs in response to their ligands and could be restored by expression of wild-type XPG. These results imply that XPG facilitates and/or stabilizes the binding of XPD and the anchoring of CAK to the core TFIIH and suggest that inadequate interaction between XPG and TFIIH can disturb the architecture of TFIIH and its function in the transcription process, providing an explanation for some of the CS features of XP-G/CS. This observation fits well with the hypothesis that transcriptional defects in certain classes of genes underlay most of the CS features and that CS is a transcriptional syndrome. It is noteworthy that all of the gene products responsible for XP/CS (XPB, XPD and XPG) are integrated in the XPG-TFIIH complex, implying that functional impairment of this complex is directly linked to the pathogenesis of CS combined with XP.

Molecular Pathogenesis of CS-Related Disorders

UV-Sensitive Syndrome with Complete Absence of CSB

UV-sensitive syndrome (UV^SS) is a rare autosomal recessive disorder characterized by photosensitivity and mild freckling, but without neurological abnormalities or skin tumors.[81] UV^SS cells show UV-hypersensitivity and defective TC-NER. It was suggested that UV^SS does not belong to any complementation groups of known photosensitive disorders such as XP and CS.[17,82] To identify the gene responsible for UV^SS, a microcell-mediated chromosome transfer was performed based on the functional complementation of UV-hypersensitivity.[4] It was found that one of the UV^SS cell lines, UV^S1KO, acquired UV-resistance when human chromosome 10 was transferred. Since the *CSB* gene is located on chromosome 10, it was sequenced and a homozygous null mutation was detected in UV^S1KO. These results indicate that the previous complementation analysis of UV^S1KO was erroneous. It was a surprising finding because a null mutation of the *CSB* gene would be expected to result in CS features such as severe developmental and neurological abnormalities.[83] On the other hand, no mutation in the *CSB* gene was detected in another UV^SS patient, Kps3, indicating the existence of genetic heterogeneity in UV^SS. Mutations of the *CSB* gene in CS-B patients have been catalogued (http://xpmutations.org/figureCSB.html); according to this database, it was expected that various types of mutant CSB polypeptides would be generated in CS-B patients. Consistent with this database, mutant CSB polypeptides were detected in CS-B-cells by Western blotting. Based on these findings, it was hypothesized that in the absence of wild-type CSB protein, the mutant CSB polypeptides produced in CS-B patients may have some inhibitory functions, which lead to the CS features. In contrast, no such mutant CSB polypeptides are generated in UV^S1KO and therefore, no such inhibitory processes are present in this patient. At present, the exact gain of function of truncated CSB proteins is unknown.

Recently, a 49-year-old Japanese female patient (KPSX6), who had a homozygous deletion of four nucleotides (CGAC) at 308-311 in the *CSB* gene, was reported. This mutation results in the frameshift followed by the stop codon at amino acid position 82. Western blot analysis using anti-N-terminal CSB antibody detected no CSB protein in KPSX6 cells. KPSX6 patient has developed freckling and dry skin in sun-exposed areas since her teens, but has been normal in mental development until 47 years old, when she began to develop dementia, ataxia and hearing loss, which had progressed over two years. Since UV^S1KO and CS3AM patients are 33 and 13 years old, respectively, they may develop the somatic and neurologic abnormalities in the future. On the other hand, two patients (CS548VI and CS539VI) from Reunion Islands, who showed somatic and neurologic CS-features and died at the age of 6 years and 8 years, respectively, had a homozygous deletion in the exon1 and upstream regulatory region of the *CSB* gene. None of the WT-CSB, CPFP and smaller CSB polypeptides was detected in these CS-B-cells. It is not known whether these two patients have an abnormality in the repair of Top1-associated lesions, or they may have a mutation in the separate gene that results in the CS-features.[84,85]

CS-Like Features Caused by Mutations in XPF and ERCC1

Xeroderma pigmentosum group F (XP-F) was first reported in Japan and to date most XP-F patients have been Japanese with a very small minority from other nations.[86,87] XP-F is characterized by mild sensitivity to sunlight causing a predisposition to skin cancer and almost no neurological abnormalities.[3] Despite the mild nature of the syndrome, deletion of the *Xpf* or *ERCC1* genes is lethal to knockout mice, which exhibit complex phenotypes with severe growth defects, multiple organ failure and early death. This is in contrast to the UV-specific phenotype of other NER-mutant mice such as *Xpa* (–/–) and *Xpc* (–/–).[88-90] Recently, it was reported that in a 15-year-old Afghan boy, a mutation in the *XPF* gene caused accelerated aging and was responsible for XFE (XPF-ERCC1) progeroid syndrome characterized by dwarfism, cachexia and microcephaly.[91] Meanwhile, a case of inherited ERCC1-deficiency was reported for the first time after 3 decades of screening patients with photosensitivity syndrome.[92] The patient's cells showed moderate hypersensitivity to UV light and mitomycin C, yet the clinical features were very severe and compatible with a diagnosis of cerebro-oculo-facio-skeletal syndrome (COFSS). The close phenotypic similarities of ERCC1-deficiency and XP-F are consistent with the functional association of the two proteins in a complex.

XPF forms a tight complex with ERCC1 and the XPF/ERCC1 complex functions as a structure-specific endonuclease responsible for the 5′ incision of a DNA lesion in NER. Moreover, the XPF/ERCC1 complex is involved in interstrand cross link-dependent 5′ and 3′ incisions during the interstrand crosslink repair process involving homologous recombination.[93-95] Intriguingly, beyond the nuclease activity in repair pathways, the XPF protein possesses a nuclease-independent activity that functions in TRF2-mediated telomere shortening.[96] In addition, XPF has been reported to be linked with tyrosyl DNA phosphodiesterase 1 (TDP1) and involved in the repair of topoisomerase I inhibitor-mediated DNA damage.[97] However, it is not yet known which function of XPF/ERCC1 is involved in XFE progeroid and COFSS syndromes.

Conclusions

As described in this chapter, our knowledge about the molecular mechanisms of TC-NER has been increased from recent studies. However, the exact functions of CSA and CSB in TC-NER remain to be fully understood. RNA polymerase IIO stalled at damaged DNA sites is ubiquitinated by several ubiquitin ligases such as Nedd4 and BRCA1/BARD1 and probably by the CSA complex (Fig. 3). It is unlikely that Nedd4 is involved in TC-NER. It is suggested that different ubiquitin ligases are involved in the different lysine-linked ubiquitinations of RNA polymerase IIO for different biological purposes. It would be important to know how these different ubiquitin ligase systems are regulated when RNA polymerase IIO encounters the damage, what is the biological significance of ubiquitination of the stalled RNA polymerase IIO and what other factors are ubiquitinated besides RNA polymerase IIO and CSB. In addition, other types of protein modifications could be involved in TC-NER.

By using in vivo crosslinking and chromatin immunoprecipitation, the isolation of a transcription elongation complex including RNA polymerase IIO stalled at UV-damaged sites revealed that CSA and CSB play distinguishable roles in the recruitment of TC-NER and core NER factors at the stalled RNA polymerase IIO site (Fig. 2). Interestingly, the chromatin remodeling factor p300 and nucleosomal binding protein HMGN1 were also recruited; their recruitment was dependent on the presence of CSB, or CSA and CSB, respectively. It is important to elucidate why chromatin remodeling is required for TC-NER.

The phenotype of CS may not be directly linked to a defect in TC-NER. The data obtained so far implied some causative relationship between CS phenotype and defects in transcription and transactivation (Fig. 4). However, more studies on the defective TC-NER and transcription in CS are important to understand the mechanisms preserving stability in the human genome.

Acknowledgements

We thank Philip C. Hanawalt, Graciela Spivak and Li Jing Tan for their critical reading of the manuscript. This work was supported by a Grant-in-aid for Scientific Research (S) from the Ministry of Education, Culture, Sports, Science and Technology (MEXT) of Japan and Solution-Oriented Research for Science and Technology (SORST), Japan Science and Technology Agency (JST).

References

1. Friedberg EC, Walker GC, Siede W et al. DNA Repair and Mutagenesis, Second Edition. ASM Press, Washington, DC, 2006.
2. Andressoo JO, Hoeijmakers JH. Transcription-coupled repair and premature ageing. Mutat Res 2005; 577:179-194.
3. Masutani C, Kusumoto R, Yamada A et al. The XPV (xeroderma pigmentosum variant) gene encodes human DNA polymerase eta. Nature 1999; 399:700-704.
4. Horibata K, Iwamoto Y, Kuraoka I et al. Complete absence of Cockayne syndrome group B gene product gives rise to UV-sensitive syndrome but not Cockayne syndrome. Proc Natl Acad Sci USA 2004; 101:15410-15415.
5. Giglia-Mari G, Coin F, Ranish JA et al. A new, tenth subunit of TFIIH is responsible for the DNA repair syndrome trichothiodystrophy group A. Nat Genet 2004; 36:714-719.
6. Nakabayashi K, Amann D, Ren Y et al. Identification of C7orf11 (TTDN1) gene mutations and genetic heterogeneity in nonphotosensitive trichothiodystrophy. Am J Hum Genet 2005; 76:510-516.
7. Sugasawa K. UV-induced ubiquitylation of XPC complex, the UV-DDB-ubiquitin ligase complex and DNA repair. J Mol Histol 2006; 37:189-202.
8. Sugasawa K, Okuda Y, Saijo M et al. UV-induced ubiquitylation of XPC protein mediated by UV-DDB-ubiquitin ligase complex. Cell 2005; 121:387-400.
9. Evans E, Moggs JG, Hwang JR et al. Mechanism of open complex and dual incision formation by human nucleotide excision repair factors. EMBO J 1997; 16:6559-6573.
10. Evans E, Fellows J, Coffer A et al. Open complex formation around a lesion during nucleotide excision repair provides a structure for cleavage by human XPG protein. EMBO J 1997; 16:625-638.
11. Coin F, Oksenych V, Egly JM. Distinct roles for the XPB/p52 and XPD/p44 subcomplexes of TFIIH in damaged DNA opening during nucleotide excision repair. Mol Cell 2007; 26:245-256.
12. Moser J, Kool H, Giakzidis I et al. Sealing of chromosomal DNA nicks during nucleotide excision repair requires XRCC1 and DNA ligase III alpha in a cell-cycle-specific manner. Mol Cell 2007; 27:311-323.
13. Ogi T, Lehmann AR. The Y-family DNA polymerase kappa (pol kappa) functions in mammalian nucleotide-excision repair. Nat Cell Biol 2006; 8:640-642.
14. Laine JP, Egly JM. Initiation of DNA repair mediated by a stalled RNA polymerase IIO. EMBO J 2006;
15. Bohr VA, Smith CA, Okumoto DS et al. DNA repair in an active gene: removal of pyrimidine dimers from the DHFR gene of CHO cells is much more efficient than in the genome overall. Cell 1985; 40:359-369.
16. Hanawalt PC. Subpathways of nucleotide excision repair and their regulation. Oncogene 2002; 21:8949-8956.
17. Itoh T, Fujiwara Y, Ono T et al. UVs syndrome, a new general category of photosensitive disorder with defective DNA repair, is distinct from xeroderma pigmentosum variant and rodent complementation group I. Am J Hum Genet 1995; 56:1267-1276.
18. Venema J, Mullenders LH, Natarajan AT et al. The genetic defect in Cockayne syndrome is associated with a defect in repair of UV-induced DNA damage in transcriptionally active DNA. Proc Natl Acad Sci USA 1990; 87:4707-4711.
19. Henning KA, Li L, Iyer N et al. The Cockayne syndrome group A gene encodes a WD repeat protein that interacts with CSB protein and a subunit of RNA polymerase II TFIIH. Cell 1995; 82:555-564.
20. Citterio E, Van Den Boom V, Schnitzler G et al. ATP-dependent chromatin remodeling by the Cockayne syndrome B DNA repair-transcription-coupling factor. Mol Cell Biol 2000; 20:7643-7653.
21. Park JS, Marr MT, Roberts JW. E. coli Transcription repair coupling factor (Mfd protein) rescues arrested complexes by promoting forward translocation. Cell 2002; 109:757-767.
22. Selby CP, Sancar A. Human transcription-repair coupling factor CSB/ERCC6 is a DNA-stimulated ATPase but is not a helicase and does not disrupt the ternary transcription complex of stalled RNA polymerase II. J Biol Chem 1997; 272:1885-1890.
23. Fousteri M, Vermeulen W, van Zeeland AA et al. Cockayne syndrome A and B proteins differentially regulate recruitment of chromatin remodeling and repair factors to stalled RNA polymerase II in vivo. Mol Cell 2006; 23:471-482.

24. Kamiuchi S, Saijo M, Citterio E et al. Translocation of Cockayne syndrome group A protein to the nuclear matrix: possible relevance to transcription-coupled DNA repair. Proc Natl Acad Sci USA 2002; 99:201-206.
25. Saijo M, Hirai T, Ogawa A et al. Functional TFIIH is required for UV-induced translocation of CSA to the nuclear matrix. Mol Cell Biol 2007; 27:2538-2547.
26. Mortillaro MJ, Blencowe BJ, Wei X et al. A hyperphosphorylated form of the large subunit of RNA polymerase II is associated with splicing complexes and the nuclear matrix. Proc Natl Acad Sci USA 1996; 93:8253-8257.
27. Koehler DR, Hanawalt PC. Recruitment of damaged DNA to the nuclear matrix in hamster cells following ultraviolet irradiation. Nucleic Acids Res 1996; 24:2877-2884.
28. Mullenders LH, van Kesteren van Leeuwen AC, van Zeeland AA et al. Nuclear matrix associated DNA is preferentially repaired in normal human fibroblasts, exposed to a low dose of ultraviolet light but not in Cockayne's syndrome fibroblasts. Nucleic Acids Res 1988; 16:10607-10622.
29. Nakatsu Y, Asahina H, Citterio E et al. XAB2, a novel tetratricopeptide repeat protein involved in transcription-coupled DNA repair and transcription. J Biol Chem 2000; 275:34931-34937.
30. Kuraoka I, Ito S, Wada T et al. Isolation of XAB2 complex involved in pre-mRNA splicing, transcription and transcription-coupled repair. J Biol Chem 2008; 283:940-950.
31. Kalogeraki VS, Tornaletti S, Cooper PK et al. Comparative TFIIS-mediated transcript cleavage by mammalian RNA polymerase II arrested at a lesion in different transcription systems. DNA Repair (Amst) 2005; 4:1075-1087.
32. Nudler E. Transcription elongation: structural basis and mechanisms. J Mol Biol 1999; 288:1-12.
33. Groisman R, Polanowska J, Kuraoka I et al. The ubiquitin ligase activity in the DDB2 and CSA complexes is differentially regulated by the COP9 signalosome in response to DNA damage. Cell 2003; 113:357-367.
34. Bregman DB, Halaban R, van Gool AJ et al. UV-induced ubiquitination of RNA polymerase II: a novel modification deficient in Cockayne syndrome cells. Proc Natl Acad Sci USA 1996; 93:11586-11590.
35. Starita LM, Horwitz AA, Keogh MC et al. BRCA1/BARD1 ubiquitinate phosphorylated RNA polymerase II. J Biol Chem 2005; 280:24498-24505.
36. Anindya R, Aygun O, Svejstrup JQ. Damage-induced ubiquitylation of human RNA polymerase II by the ubiquitin ligase Nedd4, but not Cockayne syndrome proteins or BRCA1. Mol Cell 2007; 28:386-397.
37. Wu-Baer F, Lagrazon K, Yuan W et al. The BRCA1/BARD1 heterodimer assembles polyubiquitin chains through an unconventional linkage involving lysine residue K6 of ubiquitin. J Biol Chem 2003; 278:34743-34746.
38. Groisman R, Kuraoka I, Chevallier O et al. CSA-dependent degradation of CSB by the ubiquitin-proteasome pathway establishes a link between complementation factors of the Cockayne syndrome. Genes Dev 2006; 20:1429-1434.
39. Livingstone-Zatchej M, Meier A, Suter B et al. RNA polymerase II transcription inhibits DNA repair by photolyase in the transcribed strand of active yeast genes. Nucleic Acids Res 1997; 25:3795-3800.
40. Tremeau-Bravard A, Riedl T, Egly JM et al. Fate of RNA polymerase II stalled at a cisplatin lesion. J Biol Chem 2004; 279:7751-7759.
41. Sarker AH, Tsutakawa SE, Kostek S et al. Recognition of RNA polymerase II and transcription bubbles by XPG, CSB and TFIIH: insights for transcription-coupled repair and Cockayne syndrome. Mol Cell 2005; 20:187-198.
42. Riedl T, Hanaoka F, Egly JM. The comings and goings of nucleotide excision repair factors on damaged DNA. EMBO J 2003; 22:5293-5303.
43. Squires S, Ryan AJ, Strutt HL et al. Hypersensitivity of Cockayne's syndrome cells to camptothecin is associated with the generation of abnormally high levels of double strand breaks in nascent DNA. Cancer Res 1993; 53:2012-2019.
44. Champoux JJ. DNA topoisomerases: structure, function and mechanism. Annu Rev Biochem 2001; 70:369-413.
45. Citterio E, Rademakers S, van der Horst GT et al. Biochemical and biological characterization of wild-type and ATPase-deficient Cockayne syndrome B repair protein. J Biol Chem 1998; 273:11844-11851.
46. Beerens N, Hoeijmakers JH, Kanaar R et al. The CSB protein actively wraps DNA. J Biol Chem 2005; 280:4722-4729.
47. Selby CP, Sancar A. Cockayne syndrome group B protein enhances elongation by RNA polymerase II. Proc Natl Acad Sci USA 1997; 94:11205-11209.
48. Bradsher J, Auriol J, Proietti de Santis L et al. CSB is a component of RNA pol I transcription. Mol Cell 2002; 10:819-829.
49. Yuan X, Feng W, Imhof A et al. Activation of RNA polymerase I transcription by cockayne syndrome group B protein and histone methyltransferase G9a. Mol Cell 2007; 27:585-595.

50. Iyer N, Reagan MS, Wu KJ et al. Interactions involving the human RNA polymerase II transcription/nucleotide excision repair complex TFIIH, the nucleotide excision repair protein XPG and Cockayne syndrome group B (CSB) protein. Biochemistry 1996; 35:2157-2167.
51. Yu A, Fan HY, Liao D et al. Activation of p53 or loss of the Cockayne syndrome group B repair protein causes metaphase fragility of human U1, U2 and 5S genes. Mol Cell 2000; 5:801-810.
52. Newman JC, Bailey AD, Weiner AM. Cockayne syndrome group B protein (CSB) plays a general role in chromatin maintenance and remodeling. Proc Natl Acad Sci USA 2006; 103:9613-9618.
53. Proietti-De-Santis L, Drane P, Egly JM. Cockayne syndrome B protein regulates the transcriptional program after UV irradiation. EMBO J 2006; 25:1915-1923.
54. Rockx DA, Mason R, van Hoffen A et al. UV-induced inhibition of transcription involves repression of transcription initiation and phosphorylation of RNA polymerase II. Proc Natl Acad Sci USA 2000; 97:10503-10508.
55. Rochette-Egly C. Dynamic combinatorial networks in nuclear receptor-mediated transcription. J Biol Chem 2005; 280:32565-32568.
56. Shiomi N, Mori M, Kito S et al. Severe growth retardation and short life span of double-mutant mice lacking Xpa and exon 15 of Xpg. DNA Repair (Amst) 2005; 4:351-357.
57. Zurita M, Merino C. The transcriptional complexity of the TFIIH complex. Trends Genet 2003; 19:578-584.
58. Bohr VA, Sander M, Kraemer KH. Rare diseases provide rare insights into DNA repair pathways, TFIIH, aging and cancer center. DNA Repair (Amst) 2005; 4:293-302.
59. Drapkin R, Le Roy G, Cho H et al. Human cyclin-dependent kinase-activating kinase exists in three distinct complexes. Proc Natl Acad Sci USA 1996; 93:6488-6493.
60. Reardon JT, Ge H, Gibbs E et al. Isolation and characterization of two human transcription factor IIH (TFIIH)-related complexes: ERCC2/CAK and TFIIH. Proc Natl Acad Sci USA 1996; 93:6482-6487.
61. Coin F, De Santis LP, Nardo T et al. p8/TTD-A as a repair-specific TFIIH subunit. Mol Cell 2006; 21:215-226.
62. Coin F, Bergmann E, Tremeau-Bravard A et al. Mutations in XPB and XPD helicases found in xeroderma pigmentosum patients impair the transcription function of TFIIH. EMBO J 1999; 18:1357-1366.
63. Svejstrup JQ, Vichi P, Egly JM. The multiple roles of transcription/repair factor TFIIH. Trends Biochem Sci 1996; 21:346-350.
64. Tirode F, Busso D, Coin F et al. Reconstitution of the transcription factor TFIIH: assignment of functions for the three enzymatic subunits, XPB, XPD, and cdk7. Mol Cell 1999; 3:87-95.
65. Dubaele S, Proietti De Santis L, Bienstock RJ et al. Basal transcription defect discriminates between xeroderma pigmentosum and trichothiodystrophy in XPD patients. Mol Cell 2003; 11:1635-1646.
66. Sandrock B, Egly JM. A yeast four-hybrid system identifies Cdk-activating kinase as a regulator of the XPD helicase, a subunit of transcription factor IIH. J Biol Chem 2001; 276:35328-35333.
67. Schultz P, Fribourg S, Poterszman A et al. Molecular structure of human TFIIH. Cell 2000; 102:599-607.
68. Coin F, Marinoni JC, Rodolfo C et al. Mutations in the XPD helicase gene result in XP and TTD phenotypes, preventing interaction between XPD and the p44 subunit of TFIIH. Nat Genet 1998; 20:184-188.
69. Rochette-Egly C, Adam S, Rossignol M et al. Stimulation of RAR alpha activation function AF-1 through binding to the general transcription factor TFIIH and phosphorylation by CDK7. Cell 1997; 90:97-107.
70. Bastien J, Adam-Stitah S, Riedl T et al. TFIIH interacts with the retinoic acid receptor gamma and phosphorylates its AF-1-activating domain through cdk7. J Biol Chem 2000; 275:21896-21904.
71. Chen D, Riedl T, Washbrook E et al. Activation of estrogen receptor alpha by S118 phosphorylation involves a ligand-dependent interaction with TFIIH and participation of CDK7. Mol Cell 2000; 6:127-137.
72. Compe E, Drane P, Laurent C et al. Dysregulation of the peroxisome proliferator-activated receptor target genes by XPD mutations. Mol Cell Biol 2005; 25:6065-6076.
73. Drane P, Compe E, Catez P et al. Selective regulation of vitamin D receptor-responsive genes by TFIIH. Mol Cell 2004; 16:187-197.
74. Clarkson SG. The XPG story. Biochimie 2003; 85:1113-1121.
75. Nouspikel T, Lalle P, Leadon SA et al. A common mutational pattern in Cockayne syndrome patients from xeroderma pigmentosum group G: implications for a second XPG function. Proc Natl Acad Sci USA 1997; 94:3116-3121.
76. Emmert S, Slor H, Busch DB et al. Relationship of neurologic degeneration to genotype in three xeroderma pigmentosum group G patients. J Invest Dermatol 2002; 118:972-982.

77. Harada YN, Shiomi N, Koike M et al. Postnatal growth failure, short life span and early onset of cellular senescence and subsequent immortalization in mice lacking the xeroderma pigmentosum group G gene. Mol Cell Biol 1999; 19:2366-2372.
78. Shiomi N, Kito S, Oyama M et al. Identification of the XPG region that causes the onset of Cockayne syndrome by using Xpg mutant mice generated by the cDNA-mediated knock-in method. Mol Cell Biol 2004; 24:3712-3719.
79. Lee SK, Yu SL, Prakash L et al. Requirement of yeast RAD2, a homolog of human XPG gene, for efficient RNA polymerase II transcription. Implications for Cockayne syndrome. Cell 2002; 109:823-834.
80. Ito S, Kuraoka I, Chymkowitch P, et al. XPG stabilizes TFIIH, allowing transactivation of nuclear receptors: implications for Cockayne syndrome in XP-G/CS patients. Mol Cell 2007; 26:231-243.
81. Spivak G. UV-sensitive syndrome. Mutat Res 2005; 577:162-169.
82. Itoh T, Ono T, Yamaizumi M. A new UV-sensitive syndrome not belonging to any complementation groups of xeroderma pigmentosum or Cockayne syndrome: siblings showing biochemical characteristics of Cockayne syndrome without typical clinical manifestations. Mutat Res 1994; 314:233-248.
83. Spivak G. The many faces of Cockayne syndrome. Proc Natl Acad Sci USA 2004; 101:15273-15274.
84. Hashimoto S, Suga T, Kudo E et al. Adult-onset neurological degeneration in a patient with Cockayne syndrome and a null mutation in the CSB gene. J Invest Dermatol 2008; 128:1597-1599.
85. Laugel V, Dalloz C, Stary A et al. Deletion of 5′ sequences of the CSB gene provides insight into the pathophysiology of Cockayne syndrome. Eur J Hum Genet 2008; 16:320-327.
86. Nishigori C, Fujisawa H, Uyeno K et al. Xeroderma pigmentosum patients belonging to complementation group F and efficient liquid-holding recovery of ultraviolet damage. Photodermatol Photoimmunol Photomed 1991; 8:146-150.
87. Matsumura Y, Nishigori C, Yagi T et al. Characterization of molecular defects in xeroderma pigmentosum group F in relation to its clinically mild symptoms. Hum Mol Genet 1998; 7:969-974.
88. Tian M, Shinkura R, Shinkura N et al. Growth retardation, early death and DNA repair defects in mice deficient for the nucleotide excision repair enzyme XPF. Mol Cell Biol 2004; 24:1200-1205.
89. McWhir J, Selfridge J, Harrison DJ et al. Mice with DNA repair gene (ERCC-1) deficiency have elevated levels of p53, liver nuclear abnormalities and die before weaning. Nat Genet 1993; 5:217-224.
90. Weeda G, Donker I, de Wit J et al. Disruption of mouse ERCC1 results in a novel repair syndrome with growth failure, nuclear abnormalities and senescence. Curr Biol 1997; 7:427-439.
91. Niedernhofer LJ, Garinis GA, Raams A et al. A new progeroid syndrome reveals that genotoxic stress suppresses the somatotroph axis. Nature 2006; 444:1038-1043.
92. Jaspers NG, Raams A, Silengo MC et al. First reported patient with human ERCC1 deficiency has cerebro-oculo-facio-skeletal syndrome with a mild defect in nucleotide excision repair and severe developmental failure. Am J Hum Genet 2007; 80:457-466.
93. Kuraoka I, Kobertz WR, Ariza RR et al. Repair of an interstrand DNA cross-link initiated by ERCC1-XPF repair/recombination nuclease. J Biol Chem 2000; 275:26632-26636.
94. Niedernhofer LJ, Odijk H, Budzowska M et al. The structure-specific endonuclease Ercc1-Xpf is required to resolve DNA interstrand cross-link-induced double-strand breaks. Mol Cell Biol 2004; 24:5776-5787.
95. Fisher LA, Bessho M, Bessho T. Processing of a psoralen DNA interstrand cross-link by XPF-ERCC1 complex in vitro. J Biol Chem 2008; 283:1275-1281.
96. Zhu XD, Niedernhofer L, Kuster B et al. ERCC1/XPF removes the 3′ overhang from uncapped telomeres and represses formation of telomeric DNA-containing double minute chromosomes. Mol Cell 2003; 12:1489-1498.
97. Liu C, Zhou S, Begum S, et al. Increased expression and activity of repair genes TDP1 and XPF in nonsmall cell lung cancer. Lung Cancer 2007; 55:303-311.

CHAPTER 3

Cockayne Syndrome Group B Protein and Chromatin Structure

Maria Diget Aamann and Tinna Stevnsner*

Introduction

Fitting the large amount of DNA of a eukaryotic cell into its nucleus requires the DNA to be organized together with histones in a so-called chromatin structure. The structure strengthens the DNA to allow mitosis and meiosis. It furthermore serves as a mechanism to control access of various proteins to the genome and thereby to control gene expression, DNA replication and selected DNA repair processes. There are two conceptually different ways to regulate the access of proteins to a certain area of chromatinized DNA. The histones can either be chemically modified by acetylation, methylation, phosphorylation, ADP-ribosylation or ubiquitination, which can serve as a signal to decrease the interaction between histones and DNA, or the nucleosomes can slide or be evicted from the DNA.[1] Studies indicate that the Cockayne syndrome group B (CSB) protein may play a direct or indirect role in both processes. As discussed in more detail below, in vitro studies suggest that CSB has chromatin remodeling activity, possibly associated with ATP hydrolysis and wrapping of DNA.[2,3] In addition, the recruitment of histone acetyl transferase p300 to RNA polymerase II, stalled at a UV lesion, seems to depend on the presence of CSB[4] and the acetylation status of histones in the promoter region of certain genes is affected by CSB deficiency.[5] CSB has been shown to be involved in several cellular processes including transcription regulation, transcription coupled DNA repair (TCR) and repair of some oxidatively induced lesions. As all three processes are affected by chromatin structure it seems likely that the function of CSB could be, at least partly, connected to modulation of chromatin structure.

Biochemical Data

CSB is a member of the Swi/Snf2 protein family, which includes proteins involved in transcription, recombination and DNA repair. All the family members are characterized by the presence of a conserved domain of approximately 400 amino acids including seven motifs (I, IIa, IIb, III, IV, V and VI) similar to motifs found in DNA and RNA helicases. The motifs are therefore often referred to as helicase motifs even though no actual helicase activity has yet been associated with any of the members of the Snf2 family.[6] CSB and other Snf2 family members exhibit double strand DNA (dsDNA) dependent ATPase activity, which resides in the seven conserved motifs.[7,8] The energy released from the ATP hydrolysis is possibly associated with removal of proteins from the DNA. Such activity could be an integral part of the function of CSB, possibly by destabilizing stalled RNA polymerase II as suggested by Hanawalt et al.[6] The ATPase activity of CSB is further stimulated by a 15 bp bubble.[3,7-9] The CSB protein dimerizes through the ATPase domain and results from our laboratory suggest that the ATPase activity is mainly associated with the dimer.[10]

*Corresponding Author: Tinna Stevnsner—Danish Centre for Molecular Gerontology and Danish Aging Research Centre, Aarhus University, Denmark. Email: tvs@mb.au.dk

Molecular Mechanisms of Cockayne Syndrome, edited by Shamim I. Ahmad.

A single CSB dimer seems to wrap approximately 125 bp DNA in a left handed manner and the wrapping and unwrapping is affected by ATP binding.[2]

In addition to the ATPase activity, CSB has been shown to exhibit both strand exchange and single strand annealing activity, which, at least for the latter, is independent of ATPase hydrolysis and is inhibited by binding of an un-hydrolysable ATP analogue.[11] Strand annealing has otherwise been associated with Rad51 in homologous recombination where it stimulates invasion of the presynaptic filament into homologous dsDNA in a manner which is assisted by its paralogs Rad52 and Rad54.[12] The strand annealing activity of CSB is inhibited by the single strand binding protein RPA,[11] analogous to the inhibitory effect of RPA on WRN (Werner syndrome), BLM (Bloom syndrome) and RecQ5β strand annealing activities.[13,14] The biological role of the strand annealing and strand exchange activities of CSB is still unknown.

An in vitro study has reported ATPase independent topological changes of a plasmid as a consequence of CSB binding.[3] Another study finds that CSB introduces negative supercoils in a plasmid and shortens the contour length of the plasmid with approximately 125 bp. Both functions were dependent on ATP binding but not hydrolysis.[2] Together, the shortening and the topological change suggest that CSB wraps DNA around it in a manner very similar to a nucleosome.[2] Data from DNA binding and ATPase assays indicate that binding and hydrolysis of ATP by CSB results in DNA binding and release.[2,10] It is, however, not clear whether DNA binding is triggered upon ATP binding or hydrolysis, but a mechanism with alternating DNA binding and release in a manner, which is regulated by ATP hydrolysis, would open the possibility of CSB to track along DNA and alter its topology by wrapping of the DNA. In line with this, CSB has been reported to mediate ATPase dependent nucleosome remodeling of a mononucleosome, which did not result in core histone release.[3] Nevertheless, release of histones is not a prerequisite for chromatin remodeling to have an effect on DNA repair since chromatin remodelers, that do not release histones but only opens up the chromatin structure, have been demonstrated to stimulate base excision repair (BER).[15]

CSB and Transcription

In vivo transcription in CSB deficient cells have been reported to be reduced by 50% compared to wild type cells and in vitro transcription studies on chromatin from CSB deficient cells showed that transcription elongation was significantly decreased in the absence of CSB. Furthermore, when compared to CSB deficient cells, the transcription from chromatin in wild type cells was more resistant to the detergent Triton X-100, indicating that the transcription machinery is less tightly associated with the chromatin in CSB deficient cells compared to wild type cells.[16] Weiner et al reported that the transcription pattern of CSB deficient cells overlaps significantly with the transcription pattern of cells treated with the histone deacetylase (HDAC) inhibitor TSA, indicating that CSB could play a role in chromatin maintenance.[17] The similarity between HDAC inhibited cells and CSB deficient cells thus indicates that the role of CSB is to decrease the amount of histone acetylation. In contrast, the acetylation of histone H4 is decreased near certain promoters in CSB deficient cells before and after UV irradiation, suggesting the existence of CSB dependent histone acetyl transferase (HAT) activity as argued by Egly et al.[5] Thus, the available evidence indicates that the CSB status of a cell affects the chromatin structure, but it remains unclear whether CSB is involved in increasing or decreasing the histone acetylation. The apparent contradiction could, however, be explained by the methods employed in the two studies. Weiner et al investigated the transcription pattern by microarray analysis, whereas Egly et al investigated the acetylation status of the promoter region of a few specific genes. Therefore, the difference between the general pattern throughout the genome and the pattern for specific genes could render a possible explanation for the diverging results. This notion on regulation of specific genes is supported by a microarray study investigating the transcriptional response to oxidative stress in CSB deficient cells compared to CSB complemented cells.[18] The observation here was that CSB is needed for the specific regulation of the transcription pattern after oxidative stress,[18] supporting that different genes are regulated differently by CSB.

A recent paper from Zhou et al provides evidence for involvement of CSB in histone methylation. They found that knockdown of CSB impairs RNA polymerase I interaction with rDNA and that the interaction of the histone methyl transferase G9a with RNA polymerase I depends on CSB. Furthermore, it was shown that a functional ATPase domain is needed for CSB mediated stimulation of RNA polymerase I transcription[19] lending support to a role for CSB's ATPase activity in regulation of chromatin structure. Hence, results indicate that CSB is involved in regulation of histone acetylation and methylation and these modifications are known to be involved in transcription regulation.

DNA Repair and Chromatin

In a study by Egly et al it was elegantly shown that the incision and repair of cisplatin induced DNA lesions on both the transcribed and the nontranscribed strands of a chromatinized template were increased by two transcriptional activators Gal4-VP16 and RAR, but occurred independently of transcription. Furthermore, the increased repair was shown in vitro to be dependent on the histone acetyl transferase p300, the substrate Acetyl Coenzyme A as well as ATP and ACF, an ATP dependent chromatin remodeller.[20] These observations strongly suggest that the nucleotide excision repair (NER) is stimulated by both the remodelling of nucleosomes and the acetylation of histones.

Studies on *Saccharomyces cerevisiae* provide evidence that repair of UV lesions in both the transcribed strand and nontranscribed strand is increased by overexpression of the CSB homolog, Rad26. This suggests that presence of CSB results in a more open chromatin structure favoring repair of the transcribed as well as the nontranscribed strand.[21] In support of this notion, repair of UV-induced damage in the inactive 754 locus is found to be slightly reduced in CSB deficient cells compared to wild type cells[22] and p300 recruitment to a RNA polymerase II, stalled by UV lesions, depends on CSB.[4] Point mutations in domain II, IIa, III, V and VI of the helicase domain of CSB, impair the ability of the protein to complement CSB deficient cells for the recovery of RNA synthesis after UV irradiation.[23,24] Hence, the integrity of the ATPase activity is necessary for the function of CSB in repair of UV induced lesions supporting a hypothesis of a role for CSB in chromatin remodeling in NER. A likely hypothesis, derived from the results discussed above, is that CSB increases NER activity by remodeling the chromatin structure in an ATP dependent manner, which increases the access of the repair enzymes to the DNA. The specific stimulation of TCR compared to general genome NER could be due to the interaction between a stalled RNA polymerase II and CSB. Yet, a CSB dependent stimulation of incision of UV lesions, with a stalled RNA polymerase II, has been demonstrated on a naked DNA template[25] indicating that CSB, at least in part, has a function in NER that is not related to chromatin structure.

Chromatin structure has been shown to have an inhibitory effect on the activity of some proteins involved in BER. Uracil glycosylase (UDG) and AP-endonuclease 1 (APE1) activities have been shown to be decreased on substrates with reconstituted nucleosomes. In the same study the rotational orientation of the lesion in the nucleosome was found to be of importance and the activity of DNA polymerase β was found to be completely abolished by a nucleosome context.[26] Another study has shown that the majority of the inhibitory effect of the nucleosomes is independent of histone tails.[27] Interestingly, the inhibitory effect of nucleosomes on the activity of OGG1, APE1 and DNA polymerase β, on substrates containing 8-oxoG, is relieved by Swi/Snf mediated ATP dependent chromatin remodelling.[15] This could indicate a role for CSB's chromatin remodelling activity in its function in BER. We have demonstrated an interaction between poly (ADPribose)-polymerase (PARP) and CSB and showed that the interaction is increased after oxidative stress.[28] Since PARP plays a role in both BER and in modulation of chromatin structure, it is not clear whether the interaction between CSB and PARP reflects a direct role in BER activity, or it is related to the involvement of PARP in chromatin structure alteration. However, since a variant of CSB, which lacks ATPase activity, was found to complement 8-oxoG repair in CSB deficient cells,[24] it seems likely that the ATPase activity is not important for the role of CSB in BER. In contrast, intact ATPase activity of CSB is important for its role in response to UV irradiation as discussed above. This difference in ATPase dependency indicates that the mechanisms for CSB's involvement in BER and NER are distinct. The involvement of CSB in BER could thus be related primarily to its interaction and stimulation of proteins involved in BER and less to its role in chromatin remodeling.[4]

Conclusion

The biochemical data for CSB suggest a role for the protein in nuclesome remodeling, which is ATP dependent, whereas the cellular data suggests a role in regulation of histone acetylation. These two different possibilities are not mutually exclusive since it is known that chromatin remodeling and histone acetylation are inter-connected.[29] A nucleosome remodeling function of CSB could thus result in altered histone acetylation. In support of this notion, p300 is found to be recruited to RNA polymerase II stalled by UV induced photolesions in a CSB dependent manner.[4] Thus, the involvement of CSB in transcription and repair of UV induced lesions could, in most instances, be explained by chromatin remodeling and/or recruitment of histone modifying enzymes. The involvement of CSB in BER must, at least in part, include other activities since this function is independent of ATPase activity. This notion is supported by the fact that the stimulation of APE1 endonuclease activity by CSB was seen on a nuclesome free substrate[30] strongly arguing that the role of CSB in BER goes beyond the chromatin level.

Acknowledgements

We acknowledge support by The Danish Cancer Society, The Lundbeck Foundation and The Velux Foundation. We thank Rikke Frøhlich Hougaard for critically reading the manuscript.

References

1. Osley MA, Tsukuda T, Nickoloff JA. ATP-dependent chromatin remodeling factor and DNA damage repair. Mutat Res 2007; 618:65-80.
2. Beerens N, Hoeijmakers J, H Kanaar R et al. The CSB protein actively wraps DNA. J Biol Chem 2005; 280:4722-4729.
3. Citterio E, Van Den Boom V, Schnitzler G et al. ATP-dependent chromatin remodeling by the cockayne syndrome B DNA repair-transcription-coupling factor. Mol Cell Biol 2000; 20:7643-7653.
4. Fousteri M, Vermeulen W, van Zeeland AA et al. Cockayne syndrome A and B proteins differentially regulate recruitment of chromatin remodeling and repair factors to stalled RNA polymerase II in vivo. Mol Cell 2006; 23:471-482.
5. Proietti-De-Santis L, Drane P, Egly JM. Cockayne syndrome B protein regulates the transcriptional program after UV irradiation. EMBO J 2006; 25:1915-1923.
6. Eisen JA, Sweder KS, Hanawalt PC. Evolution of the SNF2 family of proteins: subfamilies with distinct sequences and functions. Nucleic Acids Res 1995; 23:2715-2723.
7. Christiansen M, Stevnsner T, Modin C et al. Functional consequences of mutations in the conserved SF2 motifs and posttranslational phosphorylation of the CSB protein. Nucleic Acids Res 2003; 31:963-973.
8. Citterio E, Rademakers S, van der Horst GT et al. Biochemical and biological characterization of wild-type and ATPase-deficient cockayne syndrome B repair protein. J Biol Chem 1998; 273:11844-11851.
9. Selby CP, Sancar A. Human transcription-repair coupling factor CSB/ERCC6 is a DNA-stimulated ATPase but is not a helicase and does not disrupt the ternary transcription complex of stalled RNA polymerase II. J Biol Chem 1997; 272:1885-1890.
10. Christiansen M, Thorslund T, Jochimsen B et al. The cockayne syndrome group B protein is a functional dimer. Febs J 2005; 272:4306-4314.
11. Muftuoglu M, Sharma S, Thorslund T et al. Cockayne syndrome group B protein has novel strand annealing and exchange activities. Nucleic Acids Res 2006; 34:295-304.
12. Li X, Heyer WD. Homologous recombination in DNA repair and DNA damage tolerance. Cell Res 2008; 18:99-113.
13. Garcia PL, Liu Y, Jiricny J et al. Human RECQ5beta, a protein with DNA helicase and strand-annealing activities in a single polypeptide. EMBO J 2004; 23:2882-2891.
14. Machwe A, Lozada EM, Xiao L et al. Competition between the DNA unwinding and strand pairing activities of the Werner and Bloom syndrome proteins. BMC Mol Biol 2006; 7:1.
15. Menoni H, Gasparutto D, Hamiche A et al. ATP-dependent chromatin remodeling is required for base excision repair in conventional but not in variant H2A.Bbd nucleosomes. Mol Cell Biol 2007; 27:5949-5956.

16. Balajee AS, May A, Dianov GL et al. Reduced RNA polymerase II transcription in intact and permeabilized cockayne syndrome group B-cells. Proc Natl Acad Sci USA 1997; 94:4306-4311.
17. Newman JC, Bailey AD, Weiner AM. Cockayne syndrome group B protein (CSB) plays a general role in chromatin maintenance and remodeling. Proc Natl Acad Sci USA 2006; 103:9613-9618.
18. Kyng KJ, May A, Brosh RM Jr et al. The transcriptional response after oxidative stress is defective in cockayne syndrome group B-cells. Oncogen 2003; 22:1135-1149.
19. Yuan X, Feng W, Imhof A et al. Activation of RNA polymerase I transcription by cockayne syndrome group B protein and histone methyltransferase G9a. Mol Cell 2007; 27:585-595.
20. Frit P, Kwon K, Coin F et al. Transcriptional activators stimulate DNA repair. Mol Cell 2002; 10:1391-1401.
21. Bucheli M, Sweder K. In UV-irradiated saccharomyces cerevisiae, overexpression of Swi2/Snf2 family member Rad26 increases transcription-coupled repair and repair of the nontranscribed strand. Mol Microbiol 2004; 52:1653-1663.
22. Fousteri M, van Hoffen A, Vargova H et al. Repair of DNA lesions in chromosomal DNA impact of chromatin structure and cockayne syndrome proteins. DNA Repair (Amst) 2005; 4:919-925.
23. Muftuoglu M, Selzer R, Tuo J et al. Phenotypic consequences of mutations in the conserved motifs of the putative helicase domain of the human cockayne syndrome group B gene. Gene 2002; 283:27-40.
24. Selzer RR, Nyaga S, Tuo J et al. Differential requirement for the ATPase domain of the cockayne syndrome group B gene in the processing of UV-induced DNA damage and 8-oxoguanine lesions in human cells. Nucleic Acids Res 2002; 30:782-793.
25. Laine JP, Egly JM. Initiation of DNA repair mediated by a stalled RNA polymerase IIO. EMBO J 2006; 25:387-397.
26. Beard BC, Wilson SH, Smerdon MJ. Suppressed catalytic activity of base excision repair enzymes on rotationally positioned uracil in nucleosomes. Proc Natl Acad Sci USA 2003; 100:7465-7470.
27. Beard BC, Stevenson JJ, Wilson SH et al. Base excision repair in nucleosomes lacking histone tails. DNA Repair (Amst) 2005; 4:203-209.
28. Thorslund T, von Kobbe C, Harrigan JA et al. Cooperation of the cockayne syndrome group B protein and poly(ADP-ribose) polymerase 1 in the response to oxidative stress. Mol Cell Biol 2005; 25:7625-7636.
29. Eberharter A, Ferreira R, Becker P. Dynamic chromatin: concerted nucleosome remodelling and acetylation. Biol Chem 2005; 386:745-751.
30. Wong HK, Muftuoglu M, Beck G et al. Cockayne syndrome B protein stimulates apurinic endonuclease 1 activity and protects against agents that introduce base excision repair intermediates. Nucleic Acids Res 2007; 35:4103-4113.

Chapter 4

Cell Signalling, Cell Cycle Defect and Apoptosis in Cockayne Syndrome

Kiyotsugu Yoshida*

Abstract

DNA repair serves to restore two main macromolecular functions in cells: DNA replication and transcription. Defects of either of these functions can lead to apoptotic cell death. Transcription-coupled repair (TCR) is a sub-pathway of Nucleotide Excision Repair (NER) to remove bulky DNA lesions located on the transcribed strands of active genes more rapidly than those located on the nontranscribed genomic DNA. Accumulating studies have demonstrated that Cockayne syndrome is associated with defects or mutations in proteins required for TCR. In particular, the Cockayne syndrome group B (CSB) protein plays important roles in transcription, TCR and Base Excision Repair (BER). Recent structure-function studies of the proteins indicate a process-dependent variation in the molecular mechanism employed by CSB. Other studies demonstrate that CSB exerts an anti-apoptotic effect by preventing blockage of RNA polymerase II transcription in response to DNA damage. Since cell cycle arrest and induction of apoptosis, after genotoxic stress are associated with an accumulation of active p53 and inhibition of total RNA synthesis, CSB is supported to be involved in p53-dependent apoptosis. CSB is also known as a phospho-protein. In this regard, recent studies reveal that JNK and c-Abl can phosphorylate CSB. Whereas both kinases play central roles in DNA damage-signalling pathway and control cell cycle and apoptosis, CSB may function as a signalling molecule to regulate DNA damage-induced cell cycle arrest or apoptosis upon phosphorylation.

Introduction

Cockayne syndrome (CS) is an autosomal recessive multisystem disorder characterized as a segmental progeroid syndrome. CS patients display an array of clinical symptoms that include mental and physical growth retardation, demyelination, ataxia, cerebellar atrophy and calcification in the brain, particularly in the basal ganglia, gait defects, ocular anomalies such as progressive pigmentry retinopathy, sensorineural hearing loss, dental caries, mild hypersensitivity to sunlight but with no increased risk of cancer.[1,2] The protein products of these genes are involved in different aspects of the repair of damaged DNA and it has been far from clear how all these different mutations result in the same syndrome. Two complementation groups (CSA and CSB) have been identified among CS patients.[3,4] One of the hallmarks of CS cells is their inability to resume transcription after UV irradiation,[5,6] which has been commonly ascribed to the defect that these cells display in transcription coupled repair (TCR), a sub-pathway of NER. Except for the sun sensitivity, the clinical features of CS cannot be obviously attributed to defects in DNA repair,[7] which is the case for Xeroderma pigmentosum (XP) patients. It has therefore been suggested that

*Kiyotsugu Yoshida—Department of Molecular Genetics, Medical Research Institute, Tokyo Medical and Dental University, Tokyo 113-8510, Japan. Email: yos.mgen@mri.tmd.ac.jp

Molecular Mechanisms of Cockayne Syndrome, edited by Shamim I. Ahmad.

CS proteins may have additional functions beyond their role in DNA repair, possibly an involvement in transcription following genotoxic attack.[8]

CSA encodes a WD repeat protein that apparently does not interact or cofractionate with CSB protein.[9] CSB is a 168-kDa protein that belongs to the SWI2/SNF2 family of chromatin remodelling proteins,[10] exhibits ATPase activity[11] and has conserved helicase motifs.[3] CSB can interact with the XPB/ERCC3[11] and p62 components of TFIIH,[12] RNA polymerase II (pol II),[9,13] the NER protein XPG[14] and p53.[15] XPB/ERCC3 and p62 are two of the six core subunits of the dual function transcription/repair factor TFIIH that is shared between the NER and pol II basal transcription apparatus.[16] This protein has been shown to play a role in both remodelling the chromatin structure and disrupting protein—DNA interactions.[17,18] CSB is also part of and able to stimulate the enzymatic activity of complexes containing pol I and pol II. These complexes also contain TFIIH, a basal transcription factor.[11,19] Additionally, CSB cells were shown to have a defect in transcription, both in vivo and in vitro, even in the absence of stress such as UV exposure.[20] Significantly, an adequate model, describing at which stage of transcription CSB is operational beyond its involvement in TCR, is missing.

It is clear that an impasse provided by an arrested pol II would severely affect transcription. In that sense, CS could, indeed, be considered the 'transcription syndrome'. But the arrest of transcription also provides a strong signal for a cell-death (apoptosis) pathway.[21] In this context, CS could also be characterized as a disease of excessive cell death by apoptosis; a disease that adversely affect rapidly metabolizing cells, such as neurons, which generate high levels of reactive oxygen species. The apoptosis model could also, in principle, explain the problems of stunted growth and neurological deterioration. It might also explain why CS patients are not prone to skin cancer, even with severe sunlight sensitivity; after all, dead cells do not form tumors.

Interaction of CSB with p53

The p53 tumor suppressor gene is the most commonly altered gene in cancer[22] and germline transmission of a single mutant p53 allele is frequently associated with Li-Fraumeni syndrome, a disorder characterized by a predisposition to a variety of cancers.[23] The anti-neoplastic effect of p53 occurs, at least in part, via inhibiting propagation of cells with unrepaired DNA damage by enhancing DNA repair, promoting cell cycle arrest and/or facilitating apoptosis.[24] p53 accumulates in a dose-dependent manner in cells following exposure to genotoxic agents, including UV light, through transcriptional and posttranscriptional mechanisms.[25] The induction of p53 is thought to lead to the transactivation of p53-responsive genes. In addition, p53 may have transactivation-independent functions.[24,26]

CSB protein can interact with p53,[15] suggesting that p53 may play a role in transcription-coupled NER. The carboxy-terminal domain of p53 (CTD) is sufficient for interaction with CSB and deletion of the p53 CTD prevents association of p53 with CSB.[27] Interestingly, p53 mutants from human cancer cells can bind to CSB proteins to a similar or greater extent than wild type p53, in contrast to the p53 mutants that have generally lost the sequence-specific DNA binding activity needed for transcriptional activation. In this regard, it is conceivable that mutant p53 could bind to important cellular targets that are normally occupied by wild type p53, resulting in a dominant negative phenotype in the presence of wild type p53 or a dominant oncogenic phenotype in the absence of wild type p53.[28] In agreement with this concept, cells homozygous for mutant p53 are genetically more unstable than heterozygous and wild-type cells.[29,30] However, it is not clear how the binding of p53 to CSB proteins and hence the potential modulation of NER activity, relates to the phenotype of genomic stability. At least two scenarios can be postulated: CSB might recruit p53 to sites of transcription-coupled repair, or p53 might recruit CSB to sites of DNA damage, perhaps using the ability of the p53 CTD to recognize a DNA heteroduplex[31] or single-stranded DNA.[32]

In view of the role of CSB on p53-responsive gene expression, the inability to transcribe genes in CSB cells, following genotoxic attack, raises questions concerning the role of the CSB protein in general and its relationship with p53. The recovery of RNA synthesis after genotoxic attack occurs following, at least, two distinct transcriptional pathways.[33] The first one, which is directed by p53, concerns only p53-responsive genes and does not require CSB. The second one is mediated by CSB. In UV-irradiated CSB cells, p53-responsive genes are activated and the recovery of the transcription of housekeeping genes is affected. Moreover, in CSB cells, the vitamin-D dependent response is diminished, independently of UV irradiation. In this context, p53 activator possess, in addition to its DNA binding property, some additional functions exhibited by CSB. CSB was found to play a role in remodelling the chromatin structure,[17] which could facilitate the subsequent access of the general transcriptional machinery to certain genes, such as housekeeping or VDR-targeted genes. Consequently, p53 recruits specific chromatin remodelling factors to allow the initiation of the transcription for its own targeted genes. This observation would explain why p53-targeted genes are devoid of CSB, assuming the fact that CSB could function to convert the conformation of the promoter into an accessible structure.

JNK Phosphorylation of CSB

C-jun-NH_2 kinases (JNK) are amongst the UV-activated protein kinases that play an important role in cellular stress response via the phosphorylation of c-jun, ATF2 and p53.[34-36] JNK activity in the cell is tightly controlled by both, protein kinases and protein phosphatases. Various types of stimuli activate JNK through phosphorylation by the dual specificity kinases MKK4 or MKK7.[37,38] In contrast, mitogens and stress stimuli can inactivate JNK through induction of the expression of JNK phosphatases, which include serine/threonine phosphatases, tyrosine phosphatases and dual specificity (threonine/tyrosine) phosphatases.[39-41] Phosphorylation of c-jun by JNK was shown to be important for its stability and for mediating its activities in transcription, replication and transformation.[42] Activation of stress-related protein kinases by UV irradiation involves cell surface receptors and membrane components, including src tyrosine kinases as well as Grb2-SOS and Raf-1.[43-46] JNK was shown capable of interacting with Grb2 and with p21ras, which serve as docking sites for its phosphorylation by upstream kinases.[44,47] JNK activation by UV irradiation requires the presence of nuclear DNA lesion.[48,49] In this regard, study has shown a link between improper repair activity of transcribed genes and expression/activities of AP1 and ATF family members as representatives of UV-responsive element-binding proteins.[50] JNK pathway is thus one of the critical sensors and downstream effectors of persistent DNA damage in transcribing genes.[51] The interrelation between NER and transcription has been documented by findings in which DNA damage was preferentially repaired transcribing strands of active genes.[52,53] Importantly, analyses of primary fibroblast cell lines from patients with CSB revealed poor JNK activation after UV irradiation when compared with repair-proficient, normal human fibroblast cell lines.[54] Improper JNK activation is expected to affect activities of its substrates, ATF2, ELK1, c-jun and p53, which were shown to participate in the UV response in various pathways, including changes in the rate of cell cycle distribution,[55,56] growth arrest[55] and changes in the rate of DNA synthesis[57] and DNA repair.[58]

The role of JNK in apoptosis provides an alternative biological end point by which these kinases contribute to the UV response.[59-61] However, more recent studies showed that JNK signaling can regulate apoptosis both positively and negatively, depending on the cell type, cellular context and the nature and dose of treatment.[62,63] Strong and sustained JNK activation is predominantly associated with induction or enhancement of apoptosis, whereas transient JNK activation can result in cell survival.[63,64] In line with the presumed pro-apoptotic function of sustained JNK activation, the low dose UV-induced apoptotic response in CSB cells was preceded by delayed and long-lasting JNK activation.[51] Moreover, this sustained JNK activation was accompanied by prolonged phosphorylation and activation of the endogenous JNK substrate c-jun, a component of transcription factor AP-1 that can act as a pro-apoptotic effector of JNK.[65,66] Interestingly, the impaired JNK activation found in CSB cells was limited to UV radiation and was not seen after exposure to other forms of stress, such as induced by H_2O_2 or heat shock.[54] These finding suggest that there is an impaired upstream signal transduction component which is part of the pathway required to mediate UV effects. In this regard, it is reported that different forms of stress use different cellular pathways to mediate JNK activation for UV versus heat shock in transformed mouse fibroblasts.[67] DNA lesions could mediate activation of protein kinases by recruiting respective repair gene products to the damaged site. Deficient repair activity could result from improper recruitment of the necessary NER components (i.e., ERCC6, the mutant repair gene in CSB cells), yielding impaired JNK activation. Indeed, damaged DNA failed to activate JNK in vitro assays performed with CSB proteins.[54] This impairment further suggests that a signal that is emitted from damaged DNA to activate JNK is not properly sent (or received) in CSB cells.

The role of the MAPK family members p38 and ERK in DNA-damage-induced apoptosis in CSB cells remains unclear. Preliminary experiments using the p38 inhibitor SB203580 indicate that p38 does not play a significant role in the induction of apoptosis. Inhibition of ERK activation via the inhibitor U0126 also does not seem to affect UV-induced apoptosis in the human CSB cells, suggesting that ERK might play an auxiliary role. Finally, it remains to be established whether JNK-dependent apoptosis in CSB cells involves its transcription factor substrates c-Jun and ATF-2 and/or its target genes ATF-3 and c-fos. Interestingly, induction of ATF-3 can accelerate apoptosis in certain cell types,[68] whereas c-fos is anti-apoptotic in UV-treated mouse embryo fibroblasts.[69,70]

Functional Role for Tyrosine Phosphorylation of CRB by c-Abl

The c-Abl tyrosine kinase is a ubiquitously expressed proto-oncogene that contains SH3, SH2 and catalytic domains in its N-terminal region.[71] Contained within the C-terminus are nuclear localization motifs, a bipartite DNA-binding domain and F- and G-actin binding domains. Alternative splicing results in the expression of two c-Abl isoforms (1a and 1b), both of which are detectable in the nucleus and the cytoplasm. Recent studies show that c-Abl takes on an auto-inhibitory conformation and its activation requires posttranslational modifications such as phosphorylation and myristoylation.[72,73] Physiological functions dependent on c-Abl remain largely elusive. Certain insights have been derived from the findings that c-Abl is activated in response to DNA damage.[74] Nuclear c-Abl interacts with the DNA-dependent protein kinase (DNA-PK)/Ku complex.[75,76] Phosphorylation of c-Abl by the catalytic subunit DNA-PKcs induces c-Abl activity.[76] Other works have demonstrated that c-Abl is activated by the product of the gene ataxia telangiectasia mutated (ATM).[77,78] Activation of nuclear c-Abl by DNA damage contributes to induction of apoptosis by mechanisms in part dependent on the p53, p73 and Rad9.[79-83] Nuclear c-Abl also contributes to DNA damage-induced activation of the JNK/stress-activated protein kinase and p38 mitogen-activated protein kinase pathways.[74,84-86] In addition, the finding that c-Abl interacts with the Rad51 protein in response to DNA damage has supported a role for c-Abl in coordinating recombinational DNA repair with the induction of apoptosis.[87] Recent studies have demonstrated that c-Abl shuttles between the nucleus and the cytoplasm.[88,89] In contrast, oncogenic forms of Abl, including v-Abl and Bcr-Abl, localize exclusively in the cytoplasm and induce cellular transformation by promoting proliferation and inhibiting apoptotic cell death.[89-91] In this regard, recent studies have suggested that cytoplasmic c-Abl confers cell proliferation and survival.[92] By marked contrast, activation of nuclear c-Abl by many sources of DNA damage is associated with inhibition of cell growth and induction of apoptosis.[89,93] These findings indicate that the intracellular localization of c-Abl is important in dictating either survival or apoptotic responses. However, the molecular mechanism behind the nucleo-cytoplasmic shuttling of c-Abl was unknown until recently. Our recent studies revealed that c-Abl translocates into the nucleus in response to DNA damage or oxidative stress.[94]

Structure-function analysis of c-Abl indicates that it has a protein/protein interaction SH3 domain, which is likely to participate in interactions between c-Abl and its kinase substrates. SH3 domains bind preferentially to proline-rich motifs such as P-X-X-P.[95,96] Tyrosine phosphorylation, particularly by c-Abl tyrosine kinase, has been reported to play a role in the regulation of certain DNA repair proteins. c-Abl-mediated phosphorylation of DNA-topoisomerase I (Topo I) at Tyr268 in vitro and in cells conferred activation of the Topo I function.[97] Moreover, activation of c-Abl by treatment of cells with ionizing radiation was associated with c-Abl-dependent phosphorylation of Topo I and induction of Topo I activity.[97] Rad51 is a key element of recombinational DNA repair and its activity is regulated by phosphorylation of the tyrosine residue at position 315 by c-Abl tyrosine kinase.[98] Importantly, CSB is a phospho-protein with an N-terminal proline-rich region and as expected, a recent study demonstrated that CSB binds to c-Abl and is a substrate for c-Abl tyrosine kinase.[99] Oxidative stress stimulates c-Abl-mediated phosphorylation of CSB and this event is blocked by STI-571, a specific inhibitor of c-Abl kinase. Oxidative stress-induced c-Abl auto-phosphorylation and tyrosine phosphorylation of CSB appear to require CSB ATPase, because it was not observed in cells expressing an ATPase-deficient mutant of CSB. CSB and c-Abl colocalize in the nucleus and redistribute in the nucleolus in cells treated with H_2O_2. This colocalization is inhibited by pretreatment with STI-571. The redistribution of c-Abl and CSB may facilitate interaction between the two proteins and tyrosine phosphorylation of CSB. Finally, tyrosine-phosphorylated CSB may serve as a signal for repair proteins to localize to DNA damage and may help maintain active transcription in the nucleolus. Taken together, tyrosine phosphorylation of CSB plays an important role in regulating and/or coordinating the cellular response to genotoxic stress.

Concluding Remarks

The multisystem character of CS and the complexity of the genotype-phenotype relationship suggest that the underlying gene products are involved in basal cellular processes. Through the participation of CSB in several direct and functional protein interactions, the CS phenotype becomes sensitive to genetic variation affecting concentration, activity and structure of its many interaction partners.

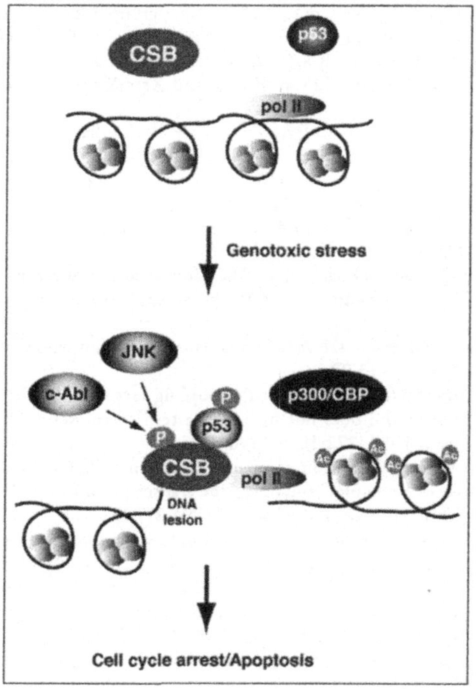

Figure 1. Current view of CSB function on transcription-coupled repair (TCR) and signaling pathways in response to DNA damage. Upon exposure to genotoxic stress, stalled RNA polymerase II (pol II) at sites of DNA lesion attracts CSB. In addition to a role in transcription-coupled repair, CSB recruits p53 to control its transcriptional activation. CSB is also phosphorylated by JNK and c-Abl, possibly allowing its involvement for cell cycle arrest or apoptosis.

The role of CSB in TCR has typically been studied using exogenous DNA-damaging agents, such as UV and cisplatin. However, UV irradiation does not penetrate the skull and therefore cannot be causative of the neurological symptoms of CS. CSB has been studied at phenotypic, cellular, subcellular and biochemical levels. Recently, several structure-function studies have aided in integrating the knowledge of these different levels. However, the question of the control of CSB itself is as yet unresolved. Given the involvement of CSB in several different processes, it can be suspected that the participation of CSB in each process is somehow inter-regulated in response to DNA damage. The inter-regulation of CSB among different cellular processes is most likely of a posttranslational manner. CSB is differentially phosphorylated in response to DNA damage (i.e., by JNK or c-Abl) and other posttranslational modifications might also be involved (Fig. 1). CSB could also be regulated, at the protein level, through changes in protein-interaction partners including p53 (Fig. 1). To investigate this, it would be interesting to map the interaction domains of CSB with the different interaction partners and to see if the different areas overlap with each other or with a putative additional CSB-binding site. A further characterization of the several functions of CSB and the underlying mechanisms employed will contribute to the understanding of not only general processes of DNA activities, but also the roles of CSB in cell signalling, cell cycle and apoptosis.

References

1. Nance MA, Berry SA. Cockayne syndrome: review of 140 cases. Am J Med Genet 1992; 42:68-84.
2. Leech RW, Brumback RA, Miller RH et al. Cockayne syndrome: clinicopathologic and tissue culture studies of affected siblings. J Neuropathol Exp Neurol 1985; 44:507-519.
3. Troelstra C, van Gool A, de Wit J et al. ERCC6, a member of a subfamily of putative helicases, is involved in Cockayne's syndrome and preferential repair of active genes. Cell 1992; 71:939-953.
4. Henning KA, Li L, Iyer N et al. The Cockayne syndrome group A gene encodes a WD repeat protein that interacts with CSB protein and a subunit of RNA polymerase II TFIIH. Cell 1995; 82:555-564.
5. Mayne LV, Lehmann AR. Failure of RNA synthesis to recover after UV irradiation: an early defect in cells from individuals with Cockayne's syndrome and Xeroderma pigmentosum. Cancer Res 1982; 42:1473-1478.
6. Mayne LV, Lehmann AR, Waters R. Excision repair in Cockayne syndrome. Mutat Res 1982; 106:179-189.
7. Lehmann AR. DNA repair-deficient diseases, Xeroderma pigmentosum, Cockayne syndrome and trichothiodystrophy. Biochimie 2003; 85:1101-1111.
8. Friedberg EC. Cockayne syndrome—a primary defect in DNA repair, transcription, both or neither? Bioessays 1996; 18:731-738.
9. van Gool AJ, Citterio E, Rademakers S et al. The Cockayne syndrome B protein, involved in transcription-coupled DNA repair, resides in an RNA polymerase II-containing complex. EMBO J 1997; 16:5955-5965.
10. Matson SW, Bean DW, George JW. DNA helicases: enzymes with essential roles in all aspects of DNA metabolism. Bioessays 1994; 16:13-22.
11. Selby CP, Sancar A. Human transcription-repair coupling factor CSB/ERCC6 is a DNA-stimulated ATPase but is not a helicase and does not disrupt the ternary transcription complex of stalled RNA polymerase II. J Biol Chem 1997; 272:1885-1890.
12. Tantin D. RNA polymerase II elongation complexes containing the Cockayne syndrome group B protein interact with a molecular complex containing the transcription factor IIH components Xeroderma pigmentosum B and p62. J Biol Chem 1998; 273:27794-27799.
13. Tantin D, Kansal A, Carey M. Recruitment of the putative transcription-repair coupling factor CSB/ERCC6 to RNA polymerase II elongation complexes. Mol Cell Biol 1997; 17:6803-6814.
14. Iyer N, Reagan MS, Wu KJ et al. Interactions involving the human RNA polymerase II transcription/nucleotide excision repair complex TFIIH, the nucleotide excision repair protein XPG and Cockayne syndrome group B (CSB) protein. Biochemistry 1996; 35:2157-2167.
15. Wang XW, Yeh H, Schaeffer L et al. p53 modulation of TFIIH-associated nucleotide excision repair activity. Nat Genet 1995; 10:188-195.
16. Marinoni JC, Roy R, Vermeulen W et al. Cloning and characterization of p52, the fifth subunit of the core of the transcription/DNA repair factor TFIIH. EMBO J 1997; 16:1093-1102.
17. Citterio E, Vermeulen W, Hoeijmakers JH. Transcriptional healing. Cell 2000; 101:447-450.
18. Beerens N, Hoeijmakers JH, Kanaar R et al. The CSB protein actively wraps DNA. J Biol Chem 2005; 280:4722-4729.

19. Bradsher J, Auriol J, Proietti de Santis L et al. CSB is a component of RNA pol I transcription. Mol Cell 2002; 10:819-829.
20. Balajee AS, May A, Dianov GL et al. Reduced RNA polymerase II transcription in intact and permeabilized Cockayne syndrome group B cells. Proc Natl Acad Sci USA 1997; 94:4306-4311.
21. Ljungman M, Zhang F. Blockage of RNA polymerase as a possible trigger for u.v. light-induced apoptosis. Oncogene 1996; 13:823-831.
22. Hollstein M, Sidransky D, Vogelstein B et al. p53 mutations in human cancers. Science 1991; 253:49-53.
23. Malkin D. p53 and the Li-Fraumeni syndrome. Biochim Biophys Acta 1994; 1198:197-213.
24. Levine AJ. p53, the cellular gatekeeper for growth and division. Cell 1997; 88:323-331.
25. Liu H, Lu ZG, Miki Y et al. Protein kinase C {delta} induces transcription of the TP53 tumor suppressor by controlling death-promoting factor Btf in the apoptotic response to DNA damage. Mol Cell Biol 2007; 24:8480-8491.
26. Ko LJ, Prives C. p53: puzzle and paradigm. Genes Dev 1996; 10:1054-1072.
27. Yu A, Fan HY, Liao D et al. Activation of p53 or loss of the Cockayne syndrome group B repair protein causes metaphase fragility of human U1, U2 and 5S genes. Mol Cell 2000; 5:801-810.
28. Dittmer D, Pati S, Zambetti G et al. Gain of function mutations in p53. Nat Genet 1993; 4:42-46.
29. Livingstone LR, White A, Sprouse J et al. Altered cell cycle arrest and gene amplification potential accompany loss of wild-type p53. Cell 1992; 70:923-935.
30. Yin Y, Tainsky MA, Bischoff FZ et al. Wild-type p53 restores cell cycle control and inhibits gene amplification in cells with mutant p53 alleles. Cell 1992; 70:937-948.
31. Lee S, Elenbaas B, Levine A et al. p53 and its 14 kDa C-terminal domain recognize primary DNA damage in the form of insertion/deletion mismatches. Cell 1995; 81:1013-1020.
32. Jayaraman J, Prives C. Activation of p53 sequence-specific DNA binding by short single strands of DNA requires the p53 C-terminus. Cell 1995; 81:1021-1029.
33. Proietti-De-Santis L, Drane P, Egly JM. Cockayne syndrome B protein regulates the transcriptional program after UV irradiation. EMBO J 2006; 25:1915-1923.
34. Derijard B, Hibi M, Wu IH et al. JNK1: a protein kinase stimulated by UV light and Ha-Ras that binds and phosphorylates the c-Jun activation domain. Cell 1994; 76:1025-1037.
35. Gupta S, Campbell D, Derijard B et al. Transcription factor ATF2 regulation by the JNK signal transduction pathway. Science 1995; 267:389-393.
36. Milne DM, Campbell LE, Campbell DG et al. p53 is phosphorylated in vitro and in vivo by an ultraviolet radiation-induced protein kinase characteristic of the c-Jun kinase, JNK1. J Biol Chem 1995; 270:5511-5518.
37. Morrison DK, Davis RJ. Regulation of MAP kinase signaling modules by scaffold proteins in mammals. Annu Rev Cell Dev Biol 2003; 19:91-118.
38. Wada T, Joza N, Cheng HY et al. MKK7 couples stress signalling to G2/M cell-cycle progression and cellular senescence. Nat Cell Biol 2004; 6:215-226.
39. Camps M, Nichols A, Arkinstall S. Dual specificity phosphatases: a gene family for control of MAP kinase function. FASEB J 2000; 14:6-16.
40. Keyse SM. Protein phosphatases and the regulation of mitogen-activated protein kinase signalling. Curr Opin Cell Biol 2000; 12:186-192.
41. Farooq A, Zhou MM. Structure and regulation of MAPK phosphatases. Cell Signal 2004; 16:769-779.
42. Smeal T, Binetruy B, Mercola DA et al. Oncogenic and transcriptional cooperation with Ha-Ras requires phosphorylation of c-Jun on serines 63 and 73. Nature 1991; 354:494-496.
43. Sachsenmaier C, Radler-Pohl A, Zinck R et al. Involvement of growth factor receptors in the mammalian UVC response. Cell 1994; 78:963-972.
44. Coffer PJ, Burgering BM, Peppelenbosch MP et al. UV activation of receptor tyrosine kinase activity. Oncogene 1995; 11:561-569.
45. Kharbanda S, Saleem A, Shafman T et al. Ionizing radiation stimulates a Grb2-mediated association of the stress-activated protein kinase with phosphatidylinositol 3-kinase. J Biol Chem 1995; 270:18871-18874.
46. Radler-Pohl A, Sachsenmaier C, Gebel S et al. UV-induced activation of AP-1 involves obligatory extranuclear steps including Raf-1 kinase. EMBO J 1993; 12:1005-1012.
47. Adler V, Pincus MR, Brandt-Rauf PW et al. Complexes of p21RAS with JUN N-terminal kinase and JUN proteins. Proc Natl Acad Sci USA 1995; 92:10585-10589.
48. Adler V, Fuchs SY, Kim J et al. jun-NH2-terminal kinase activation mediated by UV-induced DNA lesions in melanoma and fibroblast cells. Cell Growth Differ 1995; 6:1437-1446.

49. Saleem A, Datta R, Yuan ZM et al. Involvement of stress-activated protein kinase in the cellular response to 1-beta-D-arabinofuranosylcytosine and other DNA-damaging agents. Cell Growth Differ 1995; 6:1651-1658.
50. Yang YM, Ronai Z. Ultraviolet light-responsive element (TGACAACA)-binding proteins in cells of Xeroderma pigmentosum patients. Mol Carcinog 1995; 14:111-117.
51. Hamdi M, Kool J, Cornelissen-Steijger P et al. DNA damage in transcribed genes induces apoptosis via the JNK pathway and the JNK-phosphatase MKP-1. Oncogene 2005; 24:7135-7144.
52. Hanawalt PC, Donahue BA, Sweder KS. Repair and transcription. Collision or collusion? Curr Biol 1994; 4:518-521.
53. Drapkin R, Sancar A, Reinberg D. Where transcription meets repair. Cell 1994; 77:9-12.
54. Dhar V, Adler V, Lehmann A et al. Impaired jun-NH2-terminal kinase activation by ultraviolet irradiation in fibroblasts of patients with Cockayne syndrome complementation group B. Cell Growth Differ 1996; 7:841-846.
55. Kastan MB, Onyekwere O, Sidransky D et al. Participation of p53 protein in the cellular response to DNA damage. Cancer Res 1991; 51:6304-6311.
56. Yoshizumi M, Hsieh CM, Zhou F et al. The ATF site mediates downregulation of the cyclin A gene during contact inhibition in vascular endothelial cells. Mol Cell Biol 1995; 15:3266-3272.
57. Herrlich P, Rahmsdorf HJ. Transcriptional and posttranscriptional responses to DNA-damaging agents. Curr Opin Cell Biol 1994; 6:425-431.
58. Narayan S, Beard WA, Wilson SH. DNA damage-induced transcriptional activation of a human DNA polymerase beta chimeric promoter: recruitment of preinitiation complex in vitro by ATF/CREB. Biochemistry 1995; 34:73-80.
59. Xia Z, Dickens M, Raingeaud J et al. Opposing effects of ERK and JNK-p38 MAP kinases on apoptosis. Science 1995; 270:1326-1331.
60. Chen YR, Meyer CF, Tan TH. Persistent activation of c-Jun N-terminal kinase 1 (JNK1) in gamma radiation-induced apoptosis. J Biol Chem 1996; 271:631-634.
61. Yoshida K, Weichselbaum R, Kharbanda S et al. Role for Lyn tyrosine kinase as a regulator of stress-activated protein kinase activity in response to DNA damage. Mol Cell Biol 2000; 20:5370-5380.
62. Tournier C, Hess P, Yang DD et al. Requirement of JNK for stress-induced activation of the cytochrome c-mediated death pathway. Science 2000; 288:870-874.
63. Chang L, Karin M. Mammalian MAP kinase signalling cascades. Nature 2001; 410:37-40.
64. Lamb JA, Ventura JJ, Hess P et al. JunD mediates survival signaling by the JNK signal transduction pathway. Mol Cell 2003; 11:1479-1489.
65. Bossy-Wetzel E, Bakiri L, Yaniv M. Induction of apoptosis by the transcription factor c-Jun. EMBO J 1997; 16:1695-1709.
66. Behrens A, Sibilia M, Wagner EF. Amino-terminal phosphorylation of c-Jun regulates stress-induced apoptosis and cellular proliferation. Nat Genet 1999; 21:326-329.
67. Adler V, Schaffer A, Kim J et al. UV irradiation and heat shock mediate JNK activation via alternate pathways. J Biol Chem 1995; 270:26071-26077.
68. Mashima T, Udagawa S, Tsuruo T. Involvement of transcriptional repressor ATF3 in acceleration of caspase protease activation during DNA damaging agent-induced apoptosis. J Cell Physiol 2001; 188:352-358.
69. Schreiber M, Baumann B, Cotten M et al. Fos is an essential component of the mammalian UV response. EMBO J 1995; 14:5338-5349.
70. Blattner C, Kannouche P, Litfin M et al. UV-Induced stabilization of c-fos and other short-lived mRNAs. Mol Cell Biol 2000; 20:3616-3625.
71. Yoshida K. Regulation for nuclear targeting of the Abl tyrosine kinase in response to DNA damage. Adv Exp Med Biol 2007; 604:155-165.
72. Hantschel O, Nagar B, Guettler S et al. A myristoyl/phosphotyrosine switch regulates c-Abl. Cell 2003; 112:845-857.
73. Nagar B, Hantschel O, Young MA et al. Structural basis for the autoinhibition of c-Abl tyrosine kinase. Cell 2003; 112:859-871.
74. Kharbanda S, Ren R, Pandey P et al. Activation of the c-Abl tyrosine kinase in the stress response to DNA-damaging agents. Nature 1995; 376:785-788.
75. Jin S, Kharbanda S, Mayer B et al. Binding of Ku and c-Abl at the kinase homology region of DNA-dependent protein kinase catalytic subunit. J Biol Chem 1997; 272:24763-24766.
76. Kharbanda S, Pandey P, Jin S et al. Functional interaction between DNA-PK and c-Abl in response to DNA damage. Nature 1997; 386:732-735.
77. Baskaran R, Wood LD, Whitaker LL et al. Ataxia telangiectasia mutant protein activates c-Abl tyrosine kinase in response to ionizing radiation. Nature 1997; 387:516-519.

78. Shafman T, Khanna KK, Kedar P et al. Interaction between ATM protein and c-Abl in response to DNA damage. Nature 1997; 387:520-523.
79. Agami R, Blandino G, Oren M et al. Interaction of c-Abl and p73alpha and their collaboration to induce apoptosis. Nature 1999; 399:809-813.
80. Gong JG, Costanzo A, Yang HQ et al. The tyrosine kinase c-Abl regulates p73 in apoptotic response to cisplatin-induced DNA damage. Nature 1999; 399:806-809.
81. Yuan ZM, Shioya H, Ishiko T et al. p73 is regulated by tyrosine kinase c-Abl in the apoptotic response to DNA damage. Nature 1999; 399:814-817.
82. Yuan ZM, Huang Y, Fan MM et al. Genotoxic drugs induce interaction of the c-Abl tyrosine kinase and the tumor suppressor protein p53. J Biol Chem 1996; 271:26457-26460.
83. Yoshida K, Komatsu K, Wang HG et al. c-Abl tyrosine kinase regulates the human Rad9 checkpoint protein in response to DNA damage. Mol Cell Biol 2002; 22:3292-3300.
84. Kharbanda S, Pandey P, Ren R et al. c-Abl activation regulates induction of the SEK1/stress-activated protein kinase pathway in the cellular response to 1-beta-D-arabinofuranosylcytosine. J Biol Chem 1995; 270:30278-30281.
85. Kharbanda S, Pandey P, Yamauchi T et al. Activation of MEK kinase 1 by the c-Abl protein tyrosine kinase in response to DNA damage. Mol Cell Biol 2000; 20:4979-4989.
86. Pandey P, Raingeaud J, Kaneki M et al. Activation of p38 mitogen-activated protein kinase by c-Abl-dependent and -independent mechanisms. J Biol Chem 1996; 271:23775-23779.
87. Yuan ZM, Huang Y, Ishiko T et al. Regulation of Rad51 function by c-Abl in response to DNA damage. J Biol Chem 1998; 273:3799-3802.
88. Taagepera S, McDonald D, Loeb JE et al. Nuclear-cytoplasmic shuttling of C-ABL tyrosine kinase. Proc Natl Acad Sci USA 1998; 95:7457-7462.
89. Wang JY. Regulation of cell death by the Abl tyrosine kinase. Oncogene 2000; 19:5643-5650.
90. Pendergast AM. The Abl family kinases: mechanisms of regulation and signaling. Adv Cancer Res 2002; 85:51-100.
91. Wetzler M, Talpaz M, Van Etten RA et al. Subcellular localization of Bcr, Abl and Bcr-Abl proteins in normal and leukemic cells and correlation of expression with myeloid differentiation. J Clin Invest 1993; 92:1925-1939.
92. Zhu J, Wang JY. Death by Abl: a matter of location. Curr Top Dev Biol 2004; 59:165-192.
93. Kharbanda S, Yuan ZM, Weichselbaum R et al. Determination of cell fate by c-Abl activation in the response to DNA damage. Oncogene 1998; 17:3309-3318.
94. Yoshida K, Yamaguchi T, Natsume T et al. JNK phosphorylation of 14-3-3 proteins regulates nuclear targeting of c-Abl in the apoptotic response to DNA damage. Nat Cell Biol 2005; 7:278-285.
95. Kay BK, Williamson MP, Sudol M. The importance of being proline: the interaction of proline-rich motifs in signaling proteins with their cognate domains. FASEB J 2000; 14:231-241.
96. Mayer BJ. SH3 domains: complexity in moderation. J Cell Sci 2001; 114:1253-1263.
97. Yu D, Khan E, Khaleque MA et al. Phosphorylation of DNA topoisomerase I by the c-Abl tyrosine kinase confers camptothecin sensitivity. J Biol Chem 2004; 279:51851-51861.
98. Conilleau S, Takizawa Y, Tachiwana H et al. Location of tyrosine 315, a target for phosphorylation by cAbl tyrosine kinase, at the edge of the subunit-subunit interface of the human Rad51 filament. J Mol Biol 2004; 339:797-804.
99. Imam SZ, Indig FE, Cheng WH et al. Cockayne syndrome protein B interacts with and is phosphorylated by c-Abl tyrosine kinase. Nucleic Acids Res 2007; 35:4941-4951.

Chapter 5

Roles of the Cockayne Syndrome Group B Protein in Processing Oxidative DNA Damage and in Protection against Neurodegeneration

Meltem Muftuoglu and Vilhelm A. Bohr*

Abstract

Cockayne syndrome (CS) is a segmental premature aging syndrome with diverse clinical symptoms that include hypersensitivity to sunlight, severe mental and physical growth retardation, progressive neurological and retinal degeneration and skeletal abnormalities. The CS complementation group B (CSB) protein is involved in transcription coupled and global genome DNA repair, as well as in general transcription. Recent evidence suggests an additional role for the CSB protein in base excision repair (BER) of oxidative DNA lesions. Defective repair of oxidative DNA damage, from endogenous and exogenous sources in CS cells, may be the basis of the premature aging phenotype and associated neurodegeneration.

Introduction

Oxidative DNA Damage

Reactive oxygen species (ROS) are formed in living cells by normal cellular metabolism and by exogenous sources such as ionizing radiation (IR), hydrogen peroxide (H_2O_2) and genotoxic agents. These agents cause oxidative damage to DNA, resulting in the formation of base and sugar damage, strand breaks and DNA-protein crosslinks.[1] DNA base modifications are most prevalent and currently more than 20 different oxidatively modified bases have been identified.[1,2] Among them, 8-hydroxyguanine (8-oxoG) is common and mutagenic. The modified bases can mispair with adenine during DNA replication causing GC→TA transversion.[3] 8-oxoG is frequently used as a marker of oxidative levels in organisms. Although to a lesser extent than 8-oxoG, 8-hydroxyadenine (8-oxoA) also possesses premutagenic properties inducing A→G and A→C mutations in mammalian cells.[4-6] Oxidatively induced DNA damage is mainly repaired by base excision repair (BER), which involves lesion-specific DNA glycosylases in the first step of the repair process. In the absence of efficient repair, the accumulation of oxidatively induced DNA damage leads to mutagenesis and cell death/senescence. These two major oxidative stress induced DNA lesions (i.e., 8-oxoG and 8-oxoA) accumulate in Cockayne syndrome (CS) patients. It has been suggested that a decrease in the repair of oxidative DNA damage might be contributing to the pathogenesis of CS, notably premature aging (see details below). This is in line with the basis of Harman's "free

*Corresponding Author: Vilhelm A. Bohr—Laboratory of Molecular Gerontology, National Institute on Aging, NIH, 5600 Nathan Shock Drive, Baltimore, MD 21224, USA. Email: vbohr@nih.gov

Molecular Mechanisms of Cockayne Syndrome, edited by Shamim I. Ahmad.

radical theory of aging", which proposes that aging occurs through the gradual accumulation of oxidative damage to biomolecules.[7]

Base Excision Repair in Mammalian Cells

As mentioned above, BER is the major pathway for processing oxidative DNA damage (Fig. 1). BER is divided into two subpathways, either (i) a single nucleotide (short-patch) or (ii) several nucleotides incorporated (long-patch) during the repair process.[8] The first step in both short- and long-patch BER is the removal of the damaged base (e.g., 8-oxoG) by lesion-specific DNA glycosylases. These enzymes catalyze the hydrolysis of the N-glycosidic bond between the modified base and the sugar moiety to release the base and generate an apurinic/apyrimidinic (AP) site. Some glycosylases such as 8-oxoguanine glycosylase (OGG1) and the endonuclease VIII-like protein (NEIL1) are bifunctional enzymes, which also possess AP-lyase activity to cleave DNA at AP sites.[9,10] After either removal of the substrate base or incision at 3′ to AP site by bifunctional DNA glycosylase, the predominant protein, responsible for executing the next step, is AP endonuclease I

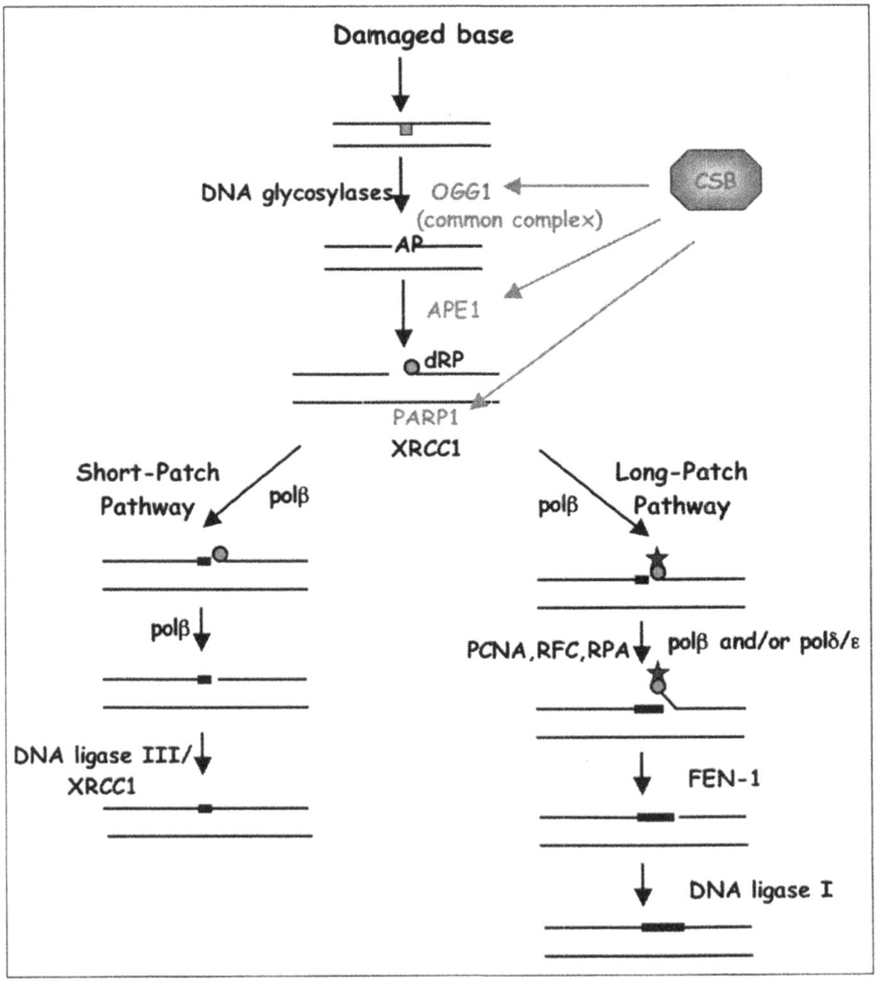

Figure 1. Base excision repair (BER) pathway. CSB are in the same complex with OGG1, interacts with APE1 and PARP-1 (see text for details).

(APE1). APE1 initiates repair of mutagenic and cytotoxic AP sites by incising the DNA backbone immediately adjacent to the lesion, creating a single strand break (SSB) with a 3′-hydroxyl terminus and a 5′-abasic residue (i.e., the deoxyribose phosphate, dRP).[11] Repair can then proceed through short or long-patch BER. The short-patch BER involves the incorporation of a single nucleotide into the gap by DNA polymerase β followed by strand ligation by DNA ligase. The long-patch BER involves incorporation of several nucleotides, typically two to seven, followed by cleavage of the resulting 5′-flap and ligation. Long-patch BER involves the proliferating cellular nuclear antigen (PCNA)-dependent polymerases, i.e., polε and polδ, replication factor C (RFC) as a loading factor, as well as the flap endonuclease 1(FEN1). SSBs are one of the intermediates in BER and are induced by alkylating agents or IR. Poly(ADP-ribosyl)polymerase-1 (PARP-1) has a role in the repair of SSBs through interaction with factors involved in BER, e.g., X-ray cross complementing 1 (XRCC1).[12] These proteins are commonly considered essential as necessary components of BER, but there are auxiliary proteins, nonessential factors, that also participate in the progression and/or regulation of BER. Many of these proteins are directly linked to human premature aging syndromes, such as Werner syndrome and CS.[12,13] Information presented in this chapter focus on the contribution of the Cockayne syndrome B (CSB) protein in the BER of oxidative DNA damage and by extrapolation supports the oxidative DNA damage theory of aging.

Cockayne Syndrome

CS is a rare genetic disorder characterized as a segmental premature aging syndrome. The clinical features of this disease include cachectic dwarfism, progressive neurological abnormalities and photosensitivity. In contrast to other photosensitive DNA repair disorders (i.e., Xeroderma Pigmentosum (XP)), CS does not show enhanced proneness to UV-induced skin cancer.[14-16] Complementation analysis has identified two genes responsible for this syndrome, *CSA* and *CSB*. Approximately 80% of CS patients have mutations in the *CSB* gene, which encodes a 168 kilodalton protein belonging to the SWI/SNF protein family.[17] CSB harbors seven helicase-like ATPase motifs like other members of this family. Although no helicase activity has been ascribed to CSB, the protein possesses DNA-dependent ATPase[18,19] and single stranded DNA annealing activities.[20] The ATPase domain is required for CSB function in chromatin remodeling [21] and modulation of negative DNA supercoiling.[22] The chromatin remodeling activity of CSB requires ATP hydrolysis, but not for the ssDNA annealing activity.[20] In fact, the ATP binding of CSB inhibits ssDNA annealing.[20] The biological relevance of this activity is not yet clear; but it may be associated with recombinational repair of DNA. Besides seven conserved helicase ATPase motifs, CSB includes an acidic region, a glycine rich region and two putative nuclear localization signal (NLS) sequences (Fig. 2). An earlier described function of CSB is in the transcription coupled repair (TCR) of UV induced DNA lesions (cyclobutane pyrimidine dimers (CPD) and 6-pyrimidine-4-pyrimidone products),[23] and recent evidence suggests an additional potential role for the CSB protein in BER of oxidative DNA lesions. Stably transformed human cell lines with site-directed mutations of the *CSB* gene in specific helicase domains were generated by replacing the highly conserved amino acids[24,25] to help understand the role of different conserved helicase motifs of CSB in the oxidative DNA damage response (see details below). However, its precise function in BER is still unclear.

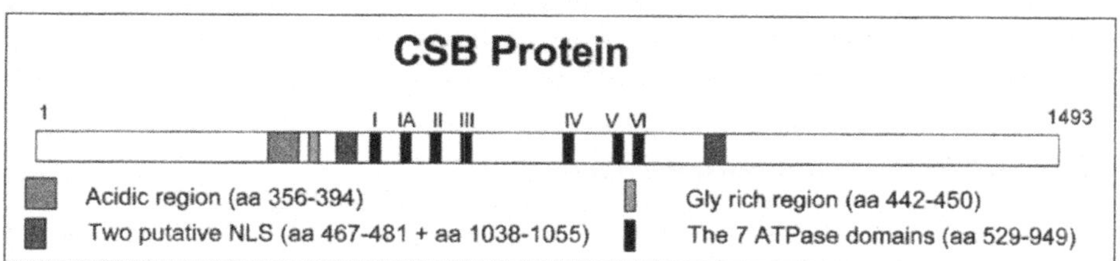

Figure 2. Schematic representation of CSB with specific domains.

Sensitivity of CSB Deficient Cells to Oxidative DNA Damaging Agents

CSB deficient cells exhibit hypersensitivity to a number of oxidative DNA damaging agents including ionizing radiation (IR), paraquat and H_2O_2.[26-29] These studies suggest that other than the UV sensitivity (a hallmark of the CS phenotype), sensitivity to oxidative lesions in CSB deficient cells might account for some of the clinical features of CS patients.

Studies on the propensity of CSB deficient cells to oxidative DNA damaging agents have been carried out with different cell types and biological systems. Many CSB studies in human cells have been conducted with primary and SV-40 transformed fibroblasts (CS1AN.S3.G2 or CS1AN) and lymphoblastoid cells derived from a patient with CS. CS1AN fibroblasts have been generated carrying a CSB allele with A→T transversions at nucleotide 1088 resulting in a stop codon at position 337 of the CSB protein.[17] Mouse models for CSB were generated by mimicking a specific truncation (as found in CSB patient CS1AN) in the mouse *CSB* gene (CSB knockout mice).[30] After the construction of the CSB knockout mice,[30] studies of their embryonic fibroblasts (MEF), keratinocytes and embryonic stem (ES) cells from isogenic mice have been undertaken. CSB studies in hamster cells are mainly performed using the UV61 cell line, which is defective in the hamster CSB homologue and which was originally used to clone the human gene for CSB.[17] Some rodent cells, including Chinese hamster ovary (CHO) cells have very limited global genome repair (GGR) of UV-induced CPD compared with human cells and thus rely primarily on TCR.[31]

Ionizing Radiation as Oxidizing Agent

CSB deficient transformed fibroblasts exposed to IR showed significant reduction in clonogenic survival.[26,27] In addition, mouse embryonic fibroblasts (MEF), embryonic stem (ES) cells and keratinocytes from CSB knockout mice show a marked reduction in survival after IR treatment.[28,29] While IR induces a variety of DNA lesions including ssDNA breaks, double-strand DNA (dsDNA) breaks and oxidative base damage, the observed hypersensitivity has been ascribed to oxidative DNA modifications such as 8-oxoG and 8-oxoA. Indeed, we have shown that after IR treatment, CSB deficient primary and transformed fibroblasts accumulate significant amounts of 8-oxoG and 8-oxoA.[27,32,33] IR doses used in these studies caused relatively few ssDNA breaks and almost no dsDNA breaks,[27] but mainly oxidative DNA base modifications. In addition, it was shown that plasmids containing 8-oxoG, are inefficiently repaired in CS patient cells, whereas strand breaks are repaired at a normal rate,[34] supporting that the increased cytotoxicity after IR is due to oxidative DNA damage. Furthermore, CSB knockout MEFs, which are sensitive to IR were also found to be hypersensitive to paraquat, which generates oxidative base lesions (see below).[28]

Another oxidative DNA base product, 5-hydroxymethyl-2′-deoxyuridine (HmdU), is formed in DNA after exposed to IR, or by attack of endogenous ROS.[35] In a recent study it has been shown that CSB deficient transformed fibroblasts are hypersensitive to HmdU, which generates cytotoxic BER intermediates/products after incorporation. It was suggested that CSB deficient cells are defective in the removal of these cytotoxic BER products, including AP sites and/or DNA SSBs.[36]

We have generated stably transformed human cell lines expressing a variety of site-directed mutations in the helicase domain of the *CSB* gene and tested for their response to oxidative agents such as IR and H_2O_2.[24,27,32] Cells mutated in helicase domains V (CSBT912V and CSBT912/913V) and VI (CSBQ942E and CSBR946A) (Fig. 2) are sensitive to IR whereas mutations in other helicase domains do not cause this effect. In addition, these mutant cells accumulated 8-oxoG and 8-oxoA lesions in genomic DNA after exposure to IR.[27,32] Thus, it is likely that CSB could be involved in the repair of oxidized purines through its helicase domains V and VI. Since the CSB protein does not appear to have catalytic helicase activity, it may depend on its ATPase activity, located to this region of the gene.

Hydrogen Peroxide as Oxidizing Agent

H_2O_2 induces a variety of DNA lesions, including oxidatively modified bases. The CSB knockout MEFs were found to be sensitive to H_2O_2,[28] although diversified results have been observed with CSB deficient transformed fibroblasts.[25] Also, no major differences in the survival of CSB deficient and proficient fibroblasts have been observed following H_2O_2 treatment.[25] Kyng et al[37] compared the expression profile of 6912 genes in CSB deficient and proficient transformed fibroblasts after H_2O_2-induced oxidative stress.[37] In the absence of exogenous stress these cells expressed a large number of genes in a similar pattern.[25] However, after oxidative stress, CSB deficient cells had a reduced expression of 122 genes compared to CSB proficient cells. The three distinct expression profiles of CSB-dependent genes, revealed by the clustering process, suggest that CSB affects different regulatory pathways including DNA repair, transcription and signal transduction. Amongst various genes studied, certain were of particular interest such as uracil-DNA glycosylase (UDG). The expression of UDG after H_2O_2 treatment was found to be affected by the CSB protein. In addition, a general deficiency in transcription, seen after oxidative stress, suggested that the CSB protein influences the regulation of transcription of certain genes involved in the repair of oxidative DNA damage.[37]

Potassium Bromate as Oxidizing Agent

Potassium bromate ($KBrO_3$) acts mostly as a one-electron oxidant that leads, almost exclusively, to the formation of 8-oxoG and formamidopyrimidine derivatives. The effect of $KBrO_3$ in OGG1-initiated BER in CSB mutant fibroblasts, which generates DNA strand breaks, was investigated by the comet assay. Elevated levels of DNA breaks, after $KBrO_3$ treatment, were observed in CSB mutant fibroblasts suggesting an inability of CSB protein to process 8-oxoG in these cells.[38] Recent studies also show that CSB deficient fibroblasts display hypersensitivity to elevated concentrations of $KBrO_3$.[39] Thus, CSB deficient cells have limited capacity in coping with elevated levels of oxidatively damaged DNA by $KBrO_3$.

Paraquat as Oxidizing Agent

Paraquat is a widely used broad spectrum herbicide toxin. Enzymatic reduction of paraquat produces paraquat radicals (PQ^+) which react with molecular oxygen to generate ROS in several tissues inflicting oxidative stress to cells and thereby causes oxidative DNA lesions.[40] Paraquat is neurotoxic and thought to contribute to pathogenesis of neurodegenerative diseases such as Parkinson's disease.[41] The MEFs from knockout CSB mice were found to be hypersensitive to paraquat.[28,29]

Methyl Methanesulfonate as Mutagen

Human fibroblasts, defective in CSB, are hypersensitive to methyl methanesulfonate (MMS).[36] MMS does not produce oxidative base lesions, but introduces BER DNA substrates/intermediates.[36] The most important biological lesion caused by MMS is N-methylation base products such as N-methyl guanine, which frequently give rise to AP sites via enhanced hydrolysis of the N-glycosylic bond or DNA glycosylase-mediated base release.[42] CSB deficient CS1AN cells, expressing an ATPase domain II mutant CSB protein (CSBE646Q), display intermediate sensitivity to MMS. These findings provide novel evidence for a direct role of CSB in the repair of BER substrates/intermediates, possibly at AP sites through interaction with APE1[36] (see details below).

Potential Role for CSB in BER of Oxidatively Damaged DNA

Biochemical assays, using recombinant CSB protein or the extracts from CSB mutant or deficient cells, support a role of CSB in the repair of oxidatively modified base lesions. Dianov et al made the first demonstration that CSB mutant cells are defective in the incision of 8-oxoG.[43] This reduced repair of 8-oxoG is associated with a down-regulation of human 8-oxoguanine DNA glycosylase (hOGG1) gene expression in CSB mutant cells,[43] leading to reduced levels of hOGG1 protein.[44] This deficiency is complemented by transfection of CSB mutant cells with the normal *CSB* gene.[43] In addition to hOGG1, two other DNA glycosylase activities, thymine glycol DNA glycosylase (hNTH) and uracil DNA glycosylase (hUDG), have been studied in CSB deficient cells. hNTH is specific for the incision of 5-hydroxycytosine and hUDG for uracil lesion. In these studies, it was found that CSB does not affect the incision activities of these DNA glycosylases.[43] 8-oxoA is another significant lesion in oxidatively damaged DNA, but the DNA glycosylase involved specifically in the repair of this lesion has not yet been identified. CSB mutant cells are defective in the cellular repair of 8-oxoA, indicating that CSB might be one of the factors important for the repair.[32] Furthermore, cell extracts from stably transformed human cell lines, with site-directed CSB mutations in various helicase domains, were used in incision assays to better understand the importance of CSB in BER. Helicase domains V and VI of CSB are important for the activity of the protein in processing of 8-oxoG lesions,[27] whereas only domain VI is involved in the repair of 8-oxoA.[32] It is possible that CSB plays an important role in the repair of oxidatively modified bases via its interaction with lesion-specific DNA glycosylases, which recognize and release the modified bases.

In humans, OGG1 exists in both isoforms, nuclear and mitochondrial (mtOGG1) which are generated by alternative splicing. The mtOGG1 protein levels are low in CSB deficient cells.[45] Stevnsner et al[45] showed a reduced 8-oxoG incision activity in both, the mitochondrial extracts of CSB deficient cells and in CSB knockout mouse liver cells, indicating a potential role for CSB in the mitochondrial repair of oxidative base damage. This activity of the CSB protein in mitochondria is specific for the repair of 8-oxoG, because CSB deficient cells have normal levels of uracil, thymine glycol or hypoxanthine incision activities.[45] Mitochondria possess an independent BER machinery, the components of which are coded by nuclear genes, resembling nuclear BER in the major molecular steps as described above. Mitochondrial BER plays a crucial role in protecting the integrity of mitochondrial DNA (mtDNA). Since mitochondria are the primary source of oxidative endogenous ROS, the accumulation of oxidative DNA damage and mutations in mtDNA may lead to mitochondrial dysfunction and consequently cell loss.[46] Certain aspects of the clinical features of CSB overlap with the phenotype associated with mitochondrial dysfunction, including severe neurological deficiencies, dysfunction in skeletal muscle and heart and premature aging. Thus, it has been suggested that CSB deficient cells accumulate mutations in mtDNA and develop mitochondrial dysfunction that contributes to the phenotype and progression of disease in CSB patients.[45]

Apart from the in vitro incision assays, the role of CSB in the repair of oxidative base damage has also been examined in vivo by the removal of formamidopyrimidine DNA glycosylase (fpg)-sensitive sites in the genome of CSB deficient cells after DNA damage induction with light-activated photosensitizers, such as acridine orange, methylene blue and RO19-8022.[45] These agents introduce 8-oxoG lesions. Fpg is an *E. coli* glycosylase that excises 8-oxoG and formamidopyrimidine lesions, followed by DNA strand cleavage. Stevnsner et al[45] demonstrated that expression of the wild type or ATPase-deficient *CSB* gene increased the repair rate of Fpg-sensitive sites in mtDNA of CSB deficient cells.[45] Furthermore, extracts from ATPase deficient CSB mutant cells (CSBE646Q) stimulate the repair of fpg-sensitive sites or the incision of 8-oxoG in mtDNA to the same extent as wild-type CSB cells, indicating that this domain or ATPase activity of the CSB protein is not required for the repair of oxidative base lesions.

The above studies suggest that CSB is directly involved in the incision process of oxidative base damage. However the mechanism by which CSB participates in the repair of 8-oxoG in both, nuclear and mitochondrial DNA, in particular in the absence of OGG1, is not yet known. Although CSB affects the function of hOGG1 (as described above), there is no direct physical interaction between CSB and hOGG1.[44] Interestingly, however, they are found in the same protein complex[44] (Fig. 1). CSB also stimulates the expression of mtOGG1, although it has not been possible to demonstrate that the CSB protein is actually present in mitochondria. It is likely that CSB protein is present at low levels in mitochondria and that the antibodies currently available may not be sensitive enough to detect it.

Recently, CSB has been shown to interact physically and/or functionally with other proteins involved in the BER pathway including PARP-1[47] and APE1[36] (Fig. 1). PARP-1 is a nuclear protein that protects the integrity of the genome by responding to oxidative DNA damage and facilitating DNA repair. CSB interacts with PARP-1 (Fig. 1) and has been found at PARP-1 sites after oxidative damage.[47] It has been suggested that PARP-1 stimulation of BER depends on CSB.[48] ATPase activity of CSB is not required for this because poly(ADP-ribosyl)ation of CSB inhibits its DNA-dependent ATPase activity.

CSB interacts with APE1 and stimulates its AP site incision activity (Fig. 1). This activation is ATP-independent.[36] It was suggested that the CSB–APE1 interaction is most critical to regions of the genome where complex DNA structures are formed, such as during transcription or replication, or at sites of recombination (telomeres), where relative concentrations of these proteins may also be higher. Thus, in addition to coordination with DNA glycosylases in endogenous base damage repair and PARP-1 in DNA damage responses, the interaction of CSB with APE1 suggests a more general function of this protein in modulating BER processes.[36]

Mouse Models for CSB

Mouse mutants for CSB provide an excellent system to study the relationship between the CSB defect and sensitivity to oxidative DNA damage in various tissues and organs and more importantly, at the level of the intact organism. CSB knockout mice mimic the CS patients phenotype in terms of the repair defect but have very mild clinical features, such as growth-retardation neurological defects and retinal degeneration.[30] To investigate the sensitivity of intact animals to oxidative DNA damage, CSB knockout mice were fed with a di(2-ethylhexy-1)phthalate (DEHP) containing diet (a potential oxidative DNA damage inducing agent in liver).[29] DEHP-treated CSB knockout mice accumulated higher level of 8-oxoG in the liver than untreated animals, whereas no significant induction of 8-oxoG in kidneys was observed. In addition, CSB knockout mice showed significant body weight loss after consumption of DEHP. These observations suggest that CSB knockout mice, at the organismal level, are sensitive to oxidative DNA damage, reflecting the observed hypersensitivity to oxidative damage in various human and mouse cultured cell types described above.[29] This is in accord with the observed tendency of CSB knockout mice to be more sensitive to the toxic or killing effects of IR than wild type mice.[28]

OGG1 knockout mice have elevated basal levels of 8-oxoG and increased spontaneous mutation frequencies.[49,50] Interestingly, the repair of 8-oxoG is not completely abolished in OGG1-deficient cells, but only reduced. It was suggested that the CSB protein might be a possible player in this back-up repair because the combined defect of OGG1 and CSB (in $CSB^{-/-}/OGG1^{-/-}$ double-knockout mice) results in a pronounced accumulation of 8-oxoG in various tissues. This also indicates an influence of CSB on protecting cells against the accumulation of endogenous oxidative damage.[49,50] The CS retinopathy is a CS-specific feature of accelerated aging. Gorgels et al[51] observed a spontaneous loss of photoreceptor cells with age in the retina of CSB knockout mice[51] and a role for oxidative DNA lesions in retinal degeneration of CS was suggested. The retina of CSB knockout mice is hypersensitive to IR suggesting that oxidative lesions may be responsible for loss of photoreceptors.[51]

Oxidative Stress, Neurodegeneration and Relations to CS Phenotype

Patients with CS display progressive neurodegeneration including demyelination of brain, delayed psychomotor development, cerebellar dysfunction, microcephaly, retinal pigmentation and mental retardation.[52,53] Exact reason(s) for the neurodegeneration of CS patients are still largely unknown. Recent pathological evidences suggest that these symptoms are due to a large deposition of protein and lipid oxidation products, including nitrotyrosine, advanced glycation end products and 4-hydroxy-2-nonenal-modified protein in the globus pallidus of the brain.[54] In addition, the deposition of 4-hydroxy-2-nonenal-modified protein was also observed in hippocampal and cerebellar dentate neurons of CS patients. Thus, oxidative stress and disturbed glutamate transport (a hallmark of neurotoxicity) may be involved in pallidal and/or cerebellar neurodegeneration of CS.[54]

It has been hypothesized that oxidative damage to DNA may be an important factor responsible for many of the neurological defects observed in CS patients. As described above, several reports demonstrate that human and mouse CS cells are sensitive to oxidative DNA damage and do not proficiently repair the oxidative DNA lesions, such as 8-oxoG. In contrast, nuclear deposition of 8-oxoG and/or altered expression of superoxide dismutase was not observed in the pallidal neurons of CS patients.[55] Although 8-oxoG lesions accumulated in various tissues of $CSB^{-/-}/OGG1^{-/-}$ double knockout mice,[49,50] these mice did not display any enhanced neurodegenerative phenotype.[56]

One reason might be that the neurological phenotype of CSB knockout mice is much less severe than that of CS patients[53,57,58] and thus mouse models may not reflect all aspects of the human condition. Several studies propose that accumulation of endogenous oxidative DNA damages in brain blocks transcription and that can cause the neurological symptoms of CS patients.[28,49,56,59] A particular class of endogenous oxidative DNA lesions is 8,5′-cyclopurine-2′-deoxynucleosides, which can block RNA Pol II transcription and are repaired by NER.[60] Accumulation of these lesions have been suggested as a possible cause of neurodegeneration in XP.[60] In contrast to the cyclopurine-deoxynucleotides, nonbulky oxidatively induced base lesions such as 8-oxoG and formamidopyrimidine dinucleotides are not likely to block transcription and are repaired by BER. There is emerging evidence that BER deficiencies are connected to neurodegeneration and specific neurological disorders.[12] Recently, it was shown that oxidative stress-induced apoptosis in neurons is correlated with imbalanced mitochondrial BER.[61] In addition, deficiencies in mtDNA repair may be associated with neurodegeneration.[12,62] As discussed above, CSB deficient cells develop mitochondrial dysfunction due to accumulation of mtDNA mutations and this might contribute to the neurological phenotype and progression of disease in CSB patients.

CSB protein has already been found in complex with both nuclear and mitochondrial OGG1. There are several other lesion specific DNA glycosylases, with which CSB could be associated. NEIL1 and NEIL2 are recently discovered bifunctional DNA glycosylases, which have broad substrate specificity, such as for 2,6-diamino-4-hydroxy-5-formamidopyrimidine (FapyG) and 4,6-diamino-5-formamidopyrimidine (FapyA).[63] NEIL1 is involved in repairing both nuclear and mtDNA oxidative base damage.[64] Mice lacking this enzyme develop a severe phenotype closely resembling metabolic syndrome in humans.[65]

Conclusion

Evidence accumulated so far demonstrates that CSB has additional functions outside NER and general transcription, possibly in BER. Further studies are needed to unravel the potential and specific role of CSB in BER and the role of BER in the CS phenotype. Through the participation of CSB in several different cellular processes, its function might be controlled by posttranslational modifications in response to DNA damage. CSB is phosphorylated by c-abl after H_2O_2 mediated oxidative DNA damage,[66] indicating that this interaction is important in response to oxidative stress. The ATPase activity of CSB is not required for its BER function.[36,47] Thus, a distinct role for CSB is suggested in facilitating BER and this function is likely separate from its role in NER, as the ATPase activity of the protein (and energy) is required for the repair of UV-induced DNA damage. In addition to defects in the repair of damage caused by ROS, further studies are required to better understand the biochemical contributions of CSB to the different DNA damage-response pathways and its role in preventing neurodegeneration.

Acknowledgements

Some of this work was supported by funds from the National Institute on Aging/National Institutes of Health Intramural program.

References

1. Dizdaroglu M. Substrate specificities and excision kinetics of DNA glycosylases involved in base-excision repair of oxidative DNA damage. Mutat Res 2003; 531(1-2):109-126.
2. Dizdaroglu M, Jaruga P, Birincioglu M et al. Free radical-induced damage to DNA: mechanisms and measurement. Free Radic Biol Med 2002; 32(11):1102-1115.
3. Grollman AP, Moriya M. Mutagenesis by 8-oxoguanine: an enemy within. Trends Genet 1993; 9(7):246-249.
4. Tan X, Grollman AP, Shibutani S. Comparison of the mutagenic properties of 8-oxo-7,8-dihydro-2′-deoxyadenosine and 8-oxo-7,8-dihydro-2′-deoxyguanosine DNA lesions in mammalian cells. Carcinogenesis 1999; 20(12):2287-2292.
5. Wang D, Kreutzer DA, Essigmann JM. Mutagenicity and repair of oxidative DNA damage: insights from studies using defined lesions. Mutat Res 1998; 400(1-2):99-115.
6. Wood ML, Esteve A, Morningstar ML et al. Genetic effects of oxidative DNA damage: comparative mutagenesis of 7,8-dihydro-8-oxoguanine and 7,8-dihydro-8-oxoadenine in Escherichia coli. Nucleic Acids Res 1992; 20(22):6023-6032.

7. Harman D. Aging: a theory based on free radical and radiation chemistry. J Gerontol 1956; 11(3):298-300.
8. Nilsen H, Krokan HE. Base excision repair in a network of defence and tolerance. Carcinogenesis 2001; 22(7):987-998.
9. Hazra TK, Hill JW, Izumi T et al. Multiple DNA glycosylases for repair of 8-oxoguanine and their potential in vivo functions. Prog Nucleic Acid Res Mol Biol 2001; 68:193-205.
10. Wallace SS, Bandaru V, Kathe SD et al. The enigma of endonuclease VIII. DNA Repair (Amst) 2003; 2(5):441-453.
11. Wilson DM, III, Barsky D. The major human abasic endonuclease: formation, consequences and repair of abasic lesions in DNA. Mutat Res 2001; 485(4):283-307.
12. Wilson DM, III, Bohr VA. The mechanics of base excision repair and its relationship to aging and disease. DNA Repair (Amst) 2007; 6(4):544-559.
13. Bohr VA. Deficient DNA repair in the human progeroid disorder, Werner syndrome. Mutat Res 2005; 577(1-2):252-259.
14. Cleaver JE. Cancer in Xeroderma pigmentosum and related disorders of DNA repair. Nat Rev Cancer 2005; 5(7):564-573.
15. Lehmann AR. DNA repair-deficient diseases, xeroderma pigmentosum, Cockayne syndrome and trichothiodystrophy. Biochimie 2003; 85(11):1101-1111.
16. Licht CL, Stevnsner T, Bohr VA. Cockayne syndrome group B cellular and biochemical functions. Am J Hum Genet 2003; 73(6):1217-1239.
17. Troelstra C, van GA, de WJ et al. ERCC6, a member of a subfamily of putative helicases, is involved in Cockayne's syndrome and preferential repair of active genes. Cell 1992; 71(6):939-953.
18. Citterio E, Rademakers S, van der Horst GT et al. Biochemical and biological characterization of wild-type and ATPase-deficient Cockayne syndrome B repair protein. J Biol Chem 1998; 273(19):11844-11851.
19. Selby CP, Sancar A. Human transcription-repair coupling factor CSB/ERCC6 is a DNA-stimulated ATPase but is not a helicase and does not disrupt the ternary transcription complex of stalled RNA polymerase II. J Biol Chem 1997; 272(3):1885-1890.
20. Muftuoglu M, Sharma S, Thorslund T et al. Cockayne syndrome group B protein has novel strand annealing and exchange activities. Nucleic Acids Res 2006; 34(1):295-304.
21. Citterio E, Van DB, V, Schnitzler G et al. ATP-dependent chromatin remodeling by the Cockayne syndrome B DNA repair-transcription-coupling factor. Mol Cell Biol 2000; 20(20):7643-7653.
22. Beerens N, Hoeijmakers JH, Kanaar R et al. The CSB protein actively wraps DNA. J Biol Chem 2005; 280(6):4722-4729.
23. van HA, Balajee AS, van Zeeland AA et al. Nucleotide excision repair and its interplay with transcription. Toxicology 2003; 193(1-2):79-90.
24. Muftuoglu M, Selzer R, Tuo J et al. Phenotypic consequences of mutations in the conserved motifs of the putative helicase domain of the human Cockayne syndrome group B gene. Gene 2002; 283(1-2):27-40.
25. Selzer RR, Nyaga S, Tuo J et al. Differential requirement for the ATPase domain of the Cockayne syndrome group B gene in the processing of UV-induced DNA damage and 8-oxoguanine lesions in human cells. Nucleic Acids Res 2002; 30(3):782-793.
26. Leadon SA, Cooper PK. Preferential repair of ionizing radiation-induced damage in the transcribed strand of an active human gene is defective in Cockayne syndrome. Proc Natl Acad Sci USA 1993; 90(22):10499-10503.
27. Tuo J, Muftuoglu M, Chen C et al. The Cockayne syndrome group B gene product is involved in general genome base excision repair of 8-hydroxyguanine in DNA. J Biol Chem 2001; 276(49):45772-45779.
28. de WH, de WJ, Gorgels TG et al. Cell type-specific hypersensitivity to oxidative damage in CSB and XPA mice. DNA Repair (Amst) 2003; 2(1):13-25.
29. de WH, de WJ, Andressoo JO et al. Different effects of CSA and CSB deficiency on sensitivity to oxidative DNA damage. Mol Cell Biol 2004; 24(18):7941-7948.
30. van der Horst GT, van SH, Berg RJ et al. Defective transcription-coupled repair in Cockayne syndrome B mice is associated with skin cancer predisposition. Cell 1997; 89(3):425-435.
31. Bohr VA, Smith CA, Okumoto DS et al. DNA repair in an active gene: removal of pyrimidine dimers from the DHFR gene of CHO cells is much more efficient than in the genome overall. Cell 1985; 40(2):359-369.
32. Tuo J, Jaruga P, Rodriguez H et al. The Cockayne syndrome group B gene product is involved in cellular repair of 8-hydroxyadenine in DNA. J Biol Chem 2002; 277(34):30832-30837.
33. Tuo J, Jaruga P, Rodriguez H et al. Primary fibroblasts of Cockayne syndrome patients are defective in cellular repair of 8-hydroxyguanine and 8-hydroxyadenine resulting from oxidative stress. FASEB J 2003; 17(6):668-674.

34. Spivak G, Hanawalt PC. Host cell reactivation of plasmids containing oxidative DNA lesions is defective in Cockayne syndrome but normal in UV-sensitive syndrome fibroblasts. DNA Repair (Amst) 2006; 5(1):13-22.
35. Boorstein RJ, Chiu LN, Teebor GW. A mammalian cell line deficient in activity of the DNA repair enzyme 5-hydroxymethyluracil-DNA glycosylase is resistant to the toxic effects of the thymidine analog 5-hydroxymethyl-2'-deoxyuridine. Mol Cell Biol 1992; 12(12):5536-5540.
36. Wong HK, Muftuoglu M, Beck G et al. Cockayne syndrome B protein stimulates apurinic endonuclease 1 activity and protects against agents that introduce base excision repair intermediates. Nucleic Acids Res 2007; 35(12):4103-4113.
37. Kyng KJ, May A, Brosh RM Jr et al. The transcriptional response after oxidative stress is defective in Cockayne syndrome group B-cells. Oncogene 2003; 22(8):1135-1149.
38. Mosesso P, Penna S, Pepe G et al. Potassium bromate but not X-rays cause unexpectedly elevated levels of DNA breakage similar to those induced by ultraviolet light in Cockayne syndrome (CS-B) fibroblasts. Cytogenet Genome Res 2004; 104(1-4):178-181.
39. Ropolo M, Degan P, Foresta M et al. Complementation of the oxidatively damaged DNA repair defect in Cockayne syndrome A and B-cells by Escherichia coli formamidopyrimidine DNA glycosylase. Free Radic Biol Med 2007; 42(12):1807-1817.
40. Ali S, Jain SK, Abdulla M et al. Paraquat induced DNA damage by reactive oxygen species. Biochem Mol Biol Int 1996; 39(1):63-67.
41. Bove J, Prou D, Perier C et al. Toxin-induced models of parkinson's disease. NeuroRx 2005; 2(3):484-494.
42. Wyatt MD, Pittman DL. Methylating agents and DNA repair responses: Methylated bases and sources of strand breaks. Chem Res Toxicol 2006; 19(12):1580-1594.
43. Dianov G, Bischoff C, Sunesen M et al. Repair of 8-oxoguanine in DNA is deficient in Cockayne syndrome group B-cells. Nucleic Acids Res 1999; 27(5):1365-1368.
44. Tuo J, Chen C, Zeng X et al. Functional crosstalk between hOgg1 and the helicase domain of Cockayne syndrome group B protein. DNA Repair (Amst) 2002; 1(11):913-927.
45. Stevnsner T, Nyaga S, de Souza-Pinto NC et al. Mitochondrial repair of 8-oxoguanine is deficient in Cockayne syndrome group B. Oncogene 2002; 21(57):8675-8682.
46. Ames BN, Shigenaga MK, Hagen TM. Mitochondrial decay in aging. Biochim Biophys Acta 1995; 1271(1):165-170.
47. Thorslund T, von KC, Harrigan JA et al. Cooperation of the Cockayne syndrome group B protein and poly(ADP-ribose) polymerase 1 in the response to oxidative stress. Mol Cell Biol 2005; 25(17):7625-7636.
48. Flohr C, Burkle A, Radicella JP et al. Poly(ADP-ribosyl)ation accelerates DNA repair in a pathway dependent on Cockayne syndrome B protein. Nucleic Acids Res 2003; 31(18):5332-5337.
49. Osterod M, Larsen E, Le PF et al. A global DNA repair mechanism involving the Cockayne syndrome B (CSB) gene product can prevent the in vivo accumulation of endogenous oxidative DNA base damage. Oncogene 2002; 21(54):8232-8239.
50. Trapp C, Reite K, Klungland A et al. Deficiency of the Cockayne Syndrome B (CSB) gene aggravates the genomic instability caused by endogenous oxidative DNA base damage in mice. Oncogene 2007; 26(27):4044-4048.
51. Gorgels TG, van dP, I, Brandt RM et al. Retinal degeneration and ionizing radiation hypersensitivity in a mouse model for Cockayne Syndrome. Mol Cell Biol 2007; 27(4):1433-1441.
52. Andressoo JO, Hoeijmakers JH. Transcription-coupled repair and premature ageing. Mutat Res 2005; 577(1-2):179-194.
53. Nance MA, Berry SA. Cockayne syndrome: review of 140 cases. Am J Med Genet 1992; 42(1):68-84.
54. Hayashi M, Itoh M, Araki S et al. Oxidative stress and disturbed glutamate transport in hereditary nucleotide repair disorders. J Neuropathol Exp Neurol 2001; 60(4):350-356.
55. Hayashi M, Araki S, Kohyama J et al. Oxidative nucleotide damage and superoxide dismutase expression in the brains of xeroderma pigmentosum group A and Cockayne syndrome. Brain Dev 2005; 27(1):34-38.
56. Laposa RR, Huang EJ, Cleaver JE. Increased apoptosis, p53 up-regulation and cerebellar neuronal degeneration in repair-deficient Cockayne syndrome mice. Proc Natl Acad Sci USA 2007; 104(4):1389-1394.
57. de BJ, Andressoo JO, de WJ et al. Premature aging in mice deficient in DNA repair and transcription. Science 2002; 296(5571):1276-1279.
58. Friedberg EC, Meira LB, Cheo DL. Database of mouse strains carrying targeted mutations in genes affecting cellular responses to DNA damage. Version 2. Mutat Res 1998; 407(3):217-226.

59. Pastoriza-Gallego M, Armier J, Sarasin A. Transcription through 8-oxoguanine in DNA repair-proficient and Csb/Ogg1 DNA repair-deficient mouse embryonic fibroblasts is dependent upon promoter strength and sequence context. Mutagenesis 2007; 22(5):343-351.
60. Brooks PJ. The case for 8,5′-cyclopurine-2′-deoxynucleosides as endogenous DNA lesions that cause neurodegeneration in xeroderma pigmentosum. Neuroscience 2007; 145(4):1407-1417.
61. Harrison JF, Hollensworth SB, Spitz DR et al. Oxidative stress-induced apoptosis in neurons correlates with mitochondrial DNA base excision repair pathway imbalance. Nucleic Acids Res 2005; 33(14):4660-4671.
62. Weissman L, Jo DG, Sorensen MM et al. Defective DNA base excision repair in brain from individuals with alzheimer's disease and amnestic mild cognitive impairment. Nucleic Acids Res 2007; 35(16):5545-5555.
63. Dou H, Mitra S, Hazra TK. Repair of oxidized bases in DNA bubble structures by human DNA glycosylases NEIL1 and NEIL2. J Biol Chem 2003; 278(50):49679-49684.
64. Hu J, de Souza-Pinto NC, Haraguchi K et al. Repair of formamidopyrimidines in DNA involves different glycosylases: role of the OGG1, NTH1 and NEIL1 enzymes. J Biol Chem 2005; 280(49):40544-40551.
65. Vartanian V, Lowell B, Minko IG et al. The metabolic syndrome resulting from a knockout of the NEIL1 DNA glycosylase. Proc Natl Acad Sci USA 2006; 103(6):1864-1869.
66. Imam SZ, Indig FE, Cheng WH et al. Cockayne syndrome protein B interacts with and is phosphorylated by c-Abl tyrosine kinase. Nucleic Acids Res 2007; 35(15):4941-4951.

Chapter 6

Structural Biology of Cockayne Syndrome Proteins, Their Interactions and Insights into DNA Repair Mechanisms

J. Jefferson P. Perry* and John A. Tainer

Abstract

Individuals with Cockayne syndrome (CS) display a progeria phenotype, which depending on the level of severity, can lead to a rapid neurological decline. Cells from CS individuals display defects in the repair of damaged DNA in actively transcribed genes and are sensitive to lesion-induced apoptosis. Cellular studies have aided isolation of five key genes that when mutated are known to cause CS, or CS-linked with the sunlight sensitive and skin cancer prone disorder xeroderma pigmentosum (XP). However, how mutations in these genes result in CS remains poorly understood. Here, we present recent gains in our knowledge of CS-linked gene function, through the analysis of structural information and partner interactions of their protein products. We describe the current understanding of how these vital proteins coordinate the chemical steps in DNA repair and integrate cellular signaling and response events. Significantly, these studies may aid both the development of a unified understanding of the cellular survival mechanisms utilized by the CS proteins after insults by genotoxic agents and the definition of how mutation in these proteins causes neurodegenerative disease.

Introduction

A large variety of potentially mutagenic DNA lesions can be produced upon exposure to carcinogenic compounds and ultraviolet (UV) radiation. This includes bulky base adducts that are produced from carcinogens such as benzo[a]pyrene, which is present in automobile exhaust fumes and tobacco smoke. Also included are cylcobutane pyrimidine dimers (CPD) and 6-4-pyrimidine-pyrimidone (6-4-PP) photoproducts, produced from the UV-promoted covalent cross-linking of adjacent pyrimidine bases. These different lesions distort the DNA-helix and in doing so are recognized and repaired by the versatile nucleotide excision repair (NER) pathway.[1] NER is conserved in eukaryotes and consists of an initial recognition step, followed by removal of a short single-strand of DNA that contains the lesion. This results in a single-strand gap that is subsequently repaired by a DNA polymerase, which uses the undamaged strand as a template. The initial recognition step can occur in two significantly different forms, providing two sub-pathways to NER. These are transcription-coupled repair (TCR) and global genome repair (GG-NER).[2] TCR functions to repair the actively transcribed strand of genes and at a rate of about four times that of GG-NER,[3] which repairs both the transcribed and nontranscribed strands. This increased rate of repair may be due to RNA polymerases blocked by lesions on the transcribed strand[4,5] pro-

*Corresponding Author: J. Jefferson P. Perry—Department of Molecular Biology and Skaggs Institute for Chemical Biology, The Scripps Research Institute, 10550 North Torrey Pines Road, La Jolla, CA 92037, USA. Email: jjperry@scripps.edu

Molecular Mechanisms of Cockayne Syndrome, edited by Shamim I. Ahmad.

viding a more stronger signal to the repair machinery than solely a lesion. Moreover, rapid repair of damage in the transcribed strand is imperative, as an unrepaired lesion leads to both reduced transcription of critical genes and provides a potent signal for apoptosis.[6]

Much of our understanding of NER has come from the study of cells from individuals with hereditary mutations in GG-NER or TCR genes that cause clinical phenotypes.[7] This includes the sunlight sensitive and skin cancer prone disorder xeroderma pigmentosum (XP), which can be caused by a mutation in one of seven key GG-NER genes, *XPA* through to *XPG*.[8] One further XP-related gene exists, *XPV* (variant), which encodes a DNA polymerase involved in DNA replication past a UV lesion.[8] Mutations in critical TCR genes can result in Cockayne syndrome (CS). The clinical phenotype of CS is sun sensitivity, severe physical and mental retardation and pronounced cachexia and these symptoms result in childhood mortality in the more severe cases.[8] Over ninety percent of CS cases are caused by mutations in either the *CSA* or *CSB* genes, which have specialized functions in the early recognition step of TCR. The remaining cases are caused by mutations in the GG-NER genes *XPB*, *XPD* and *XPG* that also function in the later steps of TCR. Mutations in these genes result in CS combined with XP. How mutations in the CS genes give rise to the clinical phenotype is poorly understood. Cells with mutations in the CS genes are sensitive to UV, being unable to recover RNA synthesis.[9] Yet, UV light clearly does not penetrate the skull to cause DNA damage that requires TCR. Instead, the CS phenotype might be related to the high levels of oxidative stress occurring in the central nervous system. A role for CS proteins being involved in the recognition and repair of oxidatively damaged DNA is contentious, but has support.[10-12]

Evidently, many of the cellular roles for CS proteins remain to be defined. Here, we describe the recent notable insights into CS protein function, gained from techniques determining CS protein architecture and methods defining CS interactions with partners. These studies are crucial because CS proteins are often required to form larger CS multi-component cellular machines, with their conformational changes and protein hand-offs likely controlling TCR pathway regulation. Thus, continued studies of this kind will increase our understanding of the molecular mechanisms behind TCR pathway progression and provide further insights into the CS phenotype.

CSA Protein

CSA protein is known to assist RNA polymerase II (RNAP II) at sites of blocks in transcription. CSA either aids RNAP II progression at sites of local secondary structure or displaces RNAP if DNA damage causes the block, to allow for repair. Mutations in the *CSA* gene (*CKN1/ERCC8*), which is located at 5q12-q31, are observed in approximately 20% of CS individuals. Ten different mutations have been identified and these include sequence deletions, missense and nonsense mutations and splice site variants[13-19] (Fig. 1). CSA encodes a 44 KDa protein that contains multiple WD (tryptophan and aspartate) repeats[13,20] (Fig. 1). WD-repeat proteins belong to a conserved family of proteins, consisting of 1-2% of eukaryotic polypeptides and known to typically function as part of multi-protein complexes.[21] Known roles for these WD-repeat proteins include important cellular process such as signal transduction, transcription regulation and apoptosis.[21]

The structure of CSA has not been determined, but structures of other WD-repeat containing proteins provide some insights into CSA structure and function. Nearly all WD-repeat proteins form a β-propeller fold, which consists of a circular, bladed propeller-like structure. Each blade consists of four anti-parallel β-strands and the WD-repeat region comprises three strands on one blade and one strand on the adjacent blade; the sharing of one strand between the blades is thought to aid to stability of the structure.[22] Seven WD repeats are observed in the CSA sequence,[20] indicating an architectural similarity to the most commonly observed group of WD-proteins. Several protein structures from this group have been determined, including Gβ, Tup1p, Aip1 and Ski8p.[23-27] Interestingly, the β-propeller fold in these structures typically forms a platform for protein interactions and in CSA, residues 73-172 containing WD repeats 2 and 3 are suggested to be a site for interactions.[20] CSA protein partners functioning in TCR that have been identified include transcription factors TFIIH[13,28] and TFIIS[29], XPA-binding protein 2 (XAB2)[29,30] and CSB.[13]

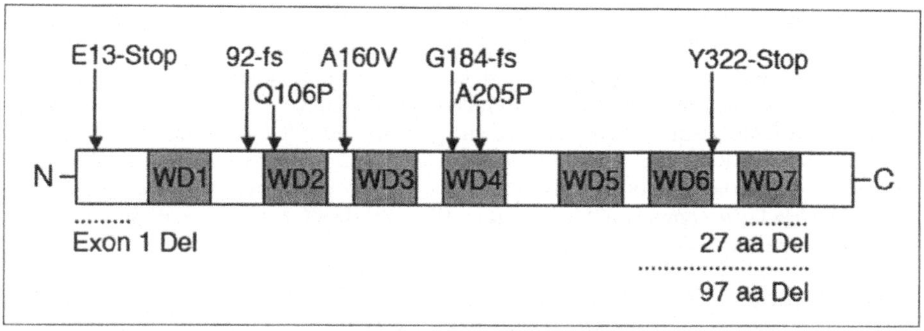

Figure 1. Schematic representation of the WD containing, 396 residue CSA protein. Mutations observed in CSA that result in CS include a large genomic deletion that removes the *CSA* exon 1 (Exon 1 Del) and a region upstream of the gene and stops protein expression.[16] Nonsense mutations include Y322 or E13 that occurred in a CS-A heterozygote patient that also had the missense A205P mutation.[13,17] Two other missense mutations are observed,[16,18] Q106P and A160V and all three of these missense mutations likely disrupt the WD motifs and hence the protein fold. 92-fs is a deletion of exon-4 region accompanied by an inversion of part of intron-4 and this causes a frameshift and stop.[16] G184-fs is a splice site acceptor mutation that adds a further 28 amino acids before a premature stop.[19] 97 and 27 amino acid C-terminal deletions are also present, marked as 97 aa Del and 27 aa Del, which are formed by splice site donor mutations that do not cause a frameshift, but the truncated protein likely has a disrupted structure.[13]

Recent studies identified that CSA belongs to a larger protein complex, which has ubiquitin ligase activity[31] and substrate specificity for CSB protein.[32] The complex contains the cullin 4A scaffold, Roc1 RING finger and the DNA damage binding protein (DDB1) adapter that interacts with CSA.[31] A second, almost identical ubiquitin complex also exists that contains another WD repeat protein, DNA damage binding protein 2 (DDB2/XPE), in place of CSA.[13,30,31] Thus, these two WD repeat proteins likely function as substrate selectors for their complexes that function opposing NER sub-pathways. Significantly, the COP9 signalosome (CSN), a regulator of cullin-based ubiquitin ligases, may interact with the CSA and DDB2 complexes.[31] CSN remains bound to the DDB2 complex and inhibits ubiquitin ligase activity, until UV irradiation causes its release. The DDB2 and CUL4A subunits of the complex are then polyubiquitinated, as are the early stage GG-NER protein, XPC and chromatin proteins.[31,33] This activity probably provides access to DNA lesions in regions of transcriptional inactivity, allowing for assembly of the GG-NER machinery. The CSA complex is CSN-free until UV radiation stimulates an interaction,[29,31] which inhibits the CSA ligase complex activity at the early stages of TCR. This initial inhibition may permit the recruitment, assembly and repair functions of the TCR pathway components. At a later stage the ligase activity of CSA complex might then allow for removal of these components when repair is complete and for resumed transcription. Notably, CSB appears to be required for the early TCR stages, before being ubiquitinated by the CSA complex and being subsequently degraded.[32] This observation lead to a hypothesis for why CSB mutations lead to CS,[32] while the complete lack of CSB protein causes UV sensitivity but not CS.[34] The mutant CSB protein may not be readily ubiquitinated by CSA and then subsequently degraded; thus bound CSB interferes with later stages in the repair process and inhibits the restart of transcription. However, potential concerns with this hypothesis include that degradation of CSB does not coincide with the measured linear kinetics of TCR over a period greater than 30 hr.[35] Also, a recent study has described a case of adult onset of neurodegeneration in a CS individual with a null mutation in the *CSB* gene.[36] An alternate study has provided a second model with some differences in functions for CSA and CSB proteins.[29] CSB was concluded to act as an initial coupling factor attracting key NER proteins and the histone acetyltransferase p300, to a stalled RNAP II.[29] While the role of CSA complex, presumably with CSB, is to recruit other required proteins likely function in TCR, including HMGN1, XAB2 and TFIIS, to the stalled RNAP II. Thus, mutations in CSA and CSB likely affect these coupling functions. However, similar to the previous model, the CSA complex also functions to prevent ubiquitination and degradation of stalled RNAP II and TCR components at a lesion site. Thus, an interesting picture is emerging, where CSA and CSB are required for the assembly of the TCR machinery and the regulation of this machinery occurs through ubiquitination. However, further studies are still required to define the precise mechanism of action of the CSA ubiquitin ligase complex and its targets in TCR and therefore how mutations in CSA give rise to a CS phenotype.

CSB Protein

Recent studies suggest that CSB protein has two key functions, recruiting TCR-pathway proteins to the site of a stalled RNA polymerase and a role in the base excision repair pathway (BER) that repairs oxidative base damage.[37] The *CSB* gene (*CKN2/ERCC6*) is located at 10q11[38] and mutations in *CSB* account for approximately eighty percent of CS cases.[14] *CSB* encodes a 1493 amino acid protein, which has sequence similarity to the SWI/SNF family of chromatin remodeling proteins[39] and functions as a dimer.[40] CSB exhibits DNA-dependent ATPase activity[41,42] and the preferred substrate is dsDNA with a shorter bubble or loop, mimicking potential DNA structures in NER and transcriptional arrest.[43]

CSB protein has seven conserved ATPase motifs,[39] which have been analyzed in homologues by structure-function studies and partially defined in CSB through mutagenesis[42,43] (Fig. 2). CSB motifs I and II are classical Walker A and B motifs, respectively, required for ATP-binding and hydrolysis.[44,45] Not surprisingly, mutations in CSB Walker A and B remove ATPase activity and also chromatin remodeling in vitro.[42,43] In vivo, Walker A and B mutations perturb recovery of RNA synthesis after UV exposure, have a failure to complement the UV-sensitivity and increased in apoptosis.[46-48] Similar defects are also observed with mutations motifs Ia and III, which are likely required for energy transduction from the ATPase site to the oligonucleotide-binding site.[44,45,49,50] Studies on motif IV have not been completed but in homologues it is required for binding dsDNA.[44,45] Motif V is expected to bind ssDNA directly and mutations in this motif and the adjacent motif VI also have similar TCR defects to mutations in the previously described motifs[49,50] and nearly

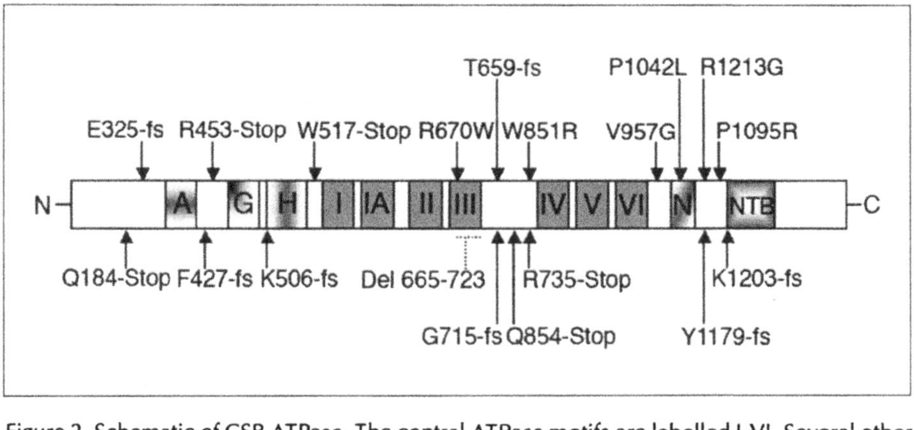

Figure 2. Schematic of CSB ATPase. The central ATPase motifs are labelled I-VI. Several other domains are present in the 1493 amino acid CSB sequence, including an acidic rich region, labelled as A, the glycine-rich stretch, G, the hydrophobic rich that contains part of bipartite NLS, H, the other part of the NLS, N and the potential nucleotide-binding motif, NTB. Mutations occur across the protein sequence and include a deletion of residues 665-723, missense mutations altering key conserved residues and likely affecting ATPase activity, in addition to nonsense mutations or frame shifts (fs) that result in truncated protein.[129,132,133]

complete inhibition of ATPase activity.[43] However, mutations in motifs V and VI differ by also having a perturbed reaction to oxidative damage, with an increased gamma-irradiation sensitivity and reduced incision of the most ubiquitous form of oxidative DNA damage, 8-hydroxy-guanine.[49] This may indicate that CSB functions in BER are linked to DNA binding rather than its ATPase activity. Notably, CSB has been observed to control the transcriptional levels of a fundamental BER protein, OGG1,[51] suggesting one potential route for CSB to affect this cellular pathway. Several other motifs are observed in CSB including, an acidic-rich region, a C-terminal Walker A motif, a glycine-rich stretch, a hydrophilic segment and a nuclear localization signal. However, mutations in these regions have not yet been observed to have affect on CSB function.[46,49,50,52]

CSB can alter both DNA conformation on binding and the arrangement of nucleosomes and also has homologous DNA strand pairing activity.[53] This indicates that CSB likely acts as a chromatin-remodeling factor[54] to either assist damage repair of active genes,[55] or to aid RNAP I- and II-mediated transcription.[56-59] CSB binds to stalled RNAP II and acting as a coupling factor to recruit downstream proteins in the repair pathway.[56,60,61] This includes of the transcription factor TFIIH, which is critically required for both GG-NER functions and TCR. Key protein-protein interactions may also mediate a potential BER role of CSB.[37] This includes that the efficient repair of oxidative base lesions occurs through an interaction of CSB and the BER protein PARP-1 (poly(ADP-ribose) polymerase 1).[62,63] CSB also interacts with BER protein apurinic/apyrimidinic endonuclease 1 (APE-1) in vivo and in vitro, which stimulates APE-1 activity in an ATP-dependent manner.[64] Moreover, CSB may be in a common complex with OGG-1, as they colocalize together after γ-irradiation.[51] However, much still remains to be defined for the functions of CSB (and its regulation by CSA). Further studies will hopefully determine which protein-partner interactions are required for the predominant roles of CSB, what are the precise mechanisms of action and how disruption of the CSB structure leads to CS phenotype.

XPB Helicase

The two helicases associated with CS, XPB and XPD, belong to the general transcription factor TFIIH.[65,66] Electron microscopy studies revealed that TFIIH forms ring-like structure that has a protrusion on one end and the central cavity of the ring is large enough to accommodate dsDNA[67,68] (Fig. 3a). The ring-core is built from XPB and XPD helicases, together with the protein partners p62, p52, p44, p32 and p8/TTDA and is sufficient to support NER functions of TFIIH.[69,70] The protrusion is the CAK complex of proteins that are required for phosphorylation of RNAP II and for transactivation of nuclear receptor genes.[71,72]

The *XPB* (*ERCC3*) gene is located at chromosome 2q21[73] and encodes a 782 amino acid protein that, similar to close homologues, unwinds DNA with a 3′-5′ polarity.[66,74,75] This helicase activity is essential for promoter DNA melting and clearance steps during RNAP II initiation of transcription and for its repair functions.[75,76] The molecular mechanisms of this dual function have begun to be revealed, through structural and biochemical studies.[77-79] The crystal structure of the archeal *Archaeoglobus fulgidus* XPB (*Af*XPB) homolog has been determined[79] (Fig. 3b). *Af*XPB shares a common core but not the N and C-termini extensions of the human protein. The *Af*XPB structure contains two RecA-like helicase domains (HD1 and HD2) that belong to helicase superfamily 2 (SNF2), as originally suggested by sequence comparison. However, other unpredicted functional regions in XPB were also discovered. This includes a small damage recognition domain (DRD) in the N-terminal region of HD1 (Fig. 3b). The DRD shares structural similarity to the mismatch recognition domain of the DNA mismatch repair protein MutS,[80] and it also recognizes damaged DNA.[79] However, XPB DRD likely recognizes NER distortions in the DNA rather than MutS-like lesions, because it is missing the critical Phe residue used in mismatch-specific interactions. Thus, the discovery of the XPB DRD region may explain how DNA damage is both located and linked to initiation of DNA unwinding, during NER steps by XPB helicase. Another discovery was the highly conserved, XPB-family specific, RED amino acid motif in domain HD1 (Fig. 3b), which mutational analysis suggests has a critical role in DNA unwinding.[79] Also observed is a thumb domain (ThM) insert in HD2, which is predicted to bind DNA independent of sequence via the

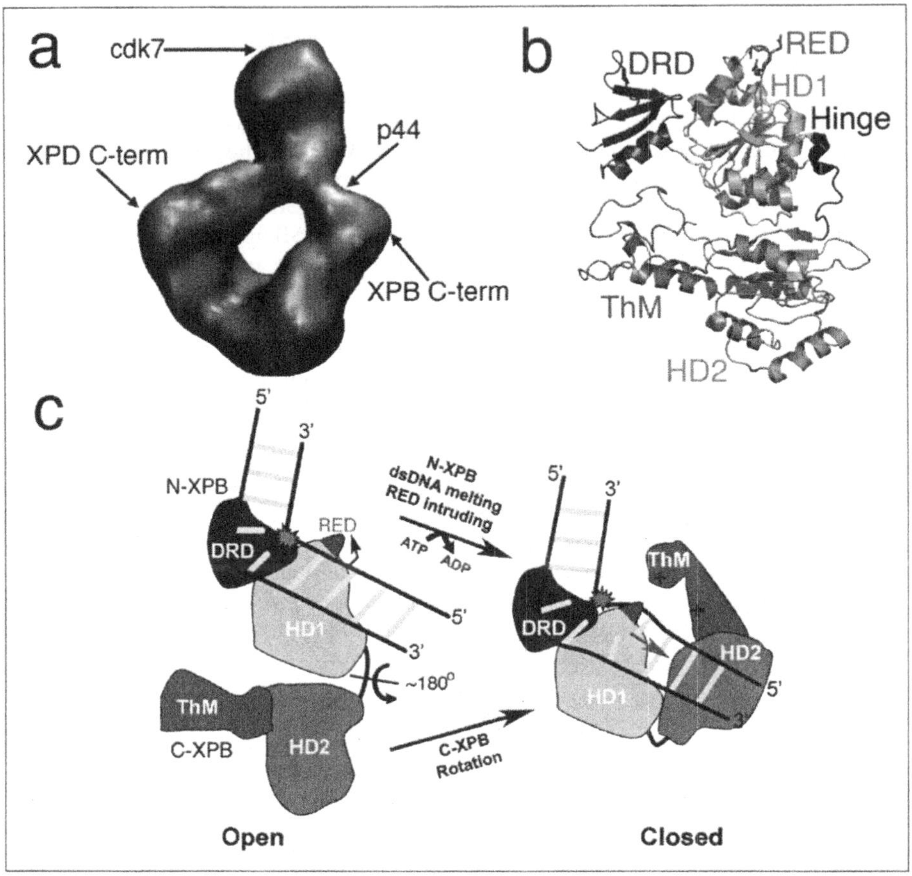

Figure 3. TFIIH and XPB structures. a) The TFIIH structure, as determined by electron microscopy,[67] has a ring-like structure with a central cavity that is large enough to accommodate dsDNA and also protrusion on one side that forms the CAK complex. The position of the cdk7 protein belonging to the CAK complex and the p44, XPB and XPD proteins that form part of the central ring were inferred by immunolabeling experiments. b) The *Af*XPB structure,[79] which contains two RecA-like helicase domains (HD1 and HD2), a small damage recognition domain (DRD), the RED motif and a thumb domain (ThM) insert. c) *Af*XPB likely changes conformation upon binding DNA, which would require a significant reorientation of domains HD1 and HD2 (panel adapted from ref. 126). This would possibly occur through the long flexible loop that connects the N- and C-terminal helicase domains. A nearly 180° change would allow XPB to wrap around the DNA and initiate unwinding.[79]

phosphodiester backbone (Fig. 3b). This is due to *Af*XPB ThM having similarity DNA polymerase ThM domains and the presence of several conserved positively charged amino acid residues at the interface between the ThM and HD2 domains.[79]

XPB likely functions through ATP hydrolysis being coupled to translocation along duplex DNA through large conformational changes in the protein, similar to that indicated for other helicases.[81,82] This is supported by the relative orientation of HD1 and HD2 in apo-*Af*XPB being[83] different to the "closed" conformation observed in crystal structures of nucleotide-bound helicases.[84] To close *Af*XPB a significant reorientation of HD1 and HD2 would have to take place (Fig. 3c) and these motions likely occur through the long flexible loop that connects the N- and C-terminal helicase domains. Thus, this structural information provides possible mechanisms for the involvement of XPB in the unwinding of duplex DNA during transcription and DNA repair. The interaction of the XPB ThM and HD2 domains with 3'-overhanging DNA may induce a reorientation of helicase domain HD2. A nearly 180° change would allow XPB to wrap around the DNA and initiate unwinding. Also, XPB DRD domain likely recognizes distorted/damaged DNA and may induce a similar conformational change during repair. In both situations this closed configuration ideally places the RED motif at the helicase active site, with its side chains intruding into the distorted DNA duplex. This likely allows the RED motif to function as a molecular wedge, to unzip the DNA when ATP hydrolysis drives XPB along the duplex DNA. However, an unconventional mechanism has also been suggest for DNA melting by XPB; where XPB potentially functions as a molecular "wrench" rotating downstream DNA relative to the fixed upstream protein-DNA interactions.[85] This would mean that the *Af*XPB structure likely represents the transcriptional mode of XPB, whereas the proposed domain reorientation is NER-specific and only occurs upon the interactions of the DRD with damaged DNA. Significantly, in either proposed mechanism the conformational change of XPB is critical for switching between the transcription and DNA repair functions of TFIIH.

Human mutations exclusively occur in the N- and C-terminal extensions of human XPB, suggesting mutation to the core region is lethal. These human XPB extensions may contribute to increased complexity of interactions and provide a greater control of function. For example, phosphorylation of Ser-751 in the C-terminal extension was reported to regulate TFIIH activity in NER.[78] Also, the interactions between XPB and partners, both within and outside of the TFIIH complex, are observed to occur in the XPB extensions and to have profound effects on the TFIIH activities in either transcription or DNA repair.[70,77] Therefore, future studies may help define how the structural and mechanistic features of XPB that are highlighted above fit into the TFIIH complex. This should include how damage recognition signals are propagated and how partner interactions coordinate the functions of XPB in NER sub-pathways.

XPD Helicase

The second helicase of the TFIIH complex, *XPD* (*ERCC2*), is located at chromosome 19q13.2-13.3 and encodes a 760 amino acid DNA helicase.[86] XPB and XPD proteins are alike in their shared similarity to SNF2 helicases, with the key family motifs being conserved in XPD (Fig. 4) and because XPD also functions in both transcription initiation and promoter escape and in NER.[87-89] However, several key differences exist between XPB and XPD proteins. For example, XPD unwinds dsDNA with the opposite 5'-3' polarity.[90] Also, XPD contains a novel iron-sulfur (Fe-S) cluster located between ATPase motifs Ia and II[91] (Fig. 4). Furthermore, mutations have been observed throughout the XPD protein and they can cause a third disorder termed trichothiodystrophy (TTD), in addition to the XP and XP-CS syndromes (Fig. 4). TTD is distinct from XP and XP-CS and its hallmark is sulfur deficient, brittle hair and other symptoms include mental retardation and reduced stature.[8]

The 5'-3' directionality of XPD indicates that its helicase activity is predominantly required to open DNA for repair. In vitro, this helicase activity is relatively weak[90] but an interaction with its partner protein in TFIIH, p44, increases activity by 10-fold.[92] This helicase activity also requires the presence of the novel Fe-S cluster,[91] which is unique to the XPD-like proteins among all characterized classes of helicases. This includes the XPD-like FANCJ helicase that causes the cancer-related Fanconi anemia disorder when mutated.. In general, Fe-S clusters are one of the most abundant and diversely employed enzymatic cofactors.[93] However, they are a rarity among nuclear proteins, since DNA glycosylases are the only other Fe-S proteins observed in the nucleus.[94,95] The Fe-S clusters are typically utilized for their ability to accept and donate electron, for their tight binding to oxygen and nitrogen atoms and for their ability to stabilize structures. Four Cys residues are conserved in eukaryotic and archeal XPD proteins and these are likely required for Fe binding; mutation in one of the Cys in the Fe-S cluster removes helicase but not ATPase activity of the archeal *Sulfolobus acidocaldarius* (*Sac*XPD) XPD homologue.[91] Moreover, *Saccharomyces cerevisiae* strains with a

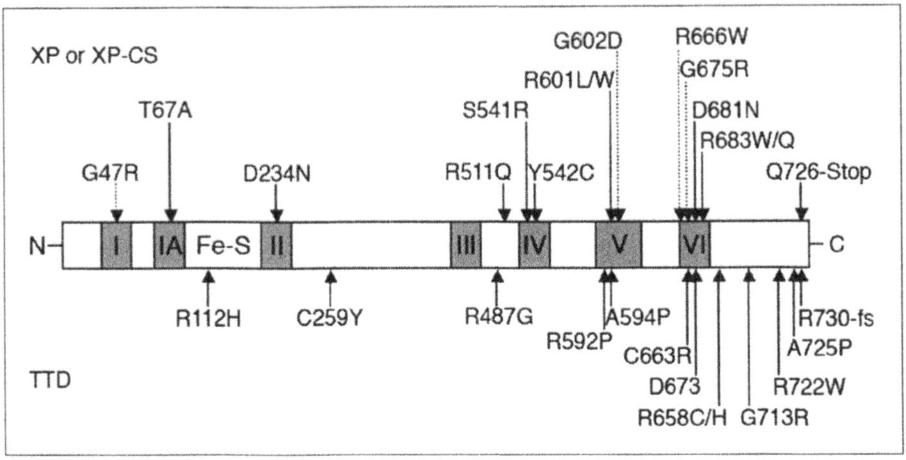

Figure 4. Schematic of XPD helicase. Three diseases, XP, XP-CS and TTD result from mutations in the 760 residue XPD helicase.[83,96,134-136] Point mutations are depicted, with TTD mutations below the domain schematic and XP mutations above, in addition to XP-CS mutations that are depicted by a dotted arrow. The XP mutations are predominantly in the ATPase domains and likely perturb ATPase and/or helicase activity. Also, many mutations cluster in the C-terminus of the protein, which is the site of interaction with the p44 subunit that stimulates XPD helicase activity.

similar Cys mutation are defective in the NER of photoproducts.[91] Notably, amino acid R112 resides in the Fe-S cluster and an R112H mutation is one of the more frequent causes of TTD.[96] The R112H mutation results in loss of XPD helicase activity in vitro and defective NER[72] and a reduction in TFIIH levels in vivo.[97] Also, mutation of the equivalent residue in *Sac*XPD removes Fe binding and helicase activity.[91] Interestingly, a potential function for the Fe-S cluster in XPD has been recently described, which helps explain how the R112H mutation affects protein function and causes the TTD phenotype: Studies on the archeal XPD protein from *Ferroplasma acidarmanus* have revealed that the intact cluster is critically required for the correct orientation of the XPD protein at the ssDNA-dsDNA junction, which is required for its unwinding activity.[98]

The three distinct disorders arising from XPD have been suggested to be due to the different transcription verses repair functions of XPD.[99] Initial support for this hypothesis came from an ATPase defective mutant in yeast XPD (Rad3). The Rad3 mutation abolished NER but the cells were still viable.[100] Later studies revealed that human XPD protein was dispensable for transcription initiation in vitro, but its presence substantially stimulates transcriptional activity.[88] More specifically, an XPD mutant that lacked ATPase activity was demonstrated to also lack helicase activity, but TFIIH containing this mutation was viable for its transcription functions.[101] However, cleavage of either side of a DNA lesion was removed by this mutation, indicating that the XPD helicase activity is required in NER,[101] most likely to allow access for the late-stage endonucleases. This also explains why XPD is relatively tolerant of mutations without affecting cell viability, since the transcriptional role of XPD is to maintain stability of the TFIIH complex. Thus, a current model is that XP and XP-CS are syndromes with repair deficiencies, since XP and XP-CS mutations are observed to remove enzymatic activity and inhibit XPD repair functions. TTD instead caused, by mutations possibly affects XPD structural stability and hence transcription.

Interestingly, many of the XP and XP-CS mutations are observed in the C-terminus of XPD, the site of interaction with the p44 partner that stimulates its activity[89] (Fig. 4). This gives further support to the hypothesis for XP and XP-CS being repair specific syndromes. However, higher resolution structural analyses would be useful in complementing the electron microscopy structure of TFIIH, to provide a much clearer understanding of mutational deficiencies. New structural information may explain how the XPD mutations are able to cause the different phenotypes, even when the mutations occur in adjacent amino acids. For example, the G602D mutation results in XP-CS, while the neighboring R601L/W mutation instead causes XP.

XPG Endonuclease

The *XPG* (*ERCC5*) gene was located at chromosome position 13q32.2-q33.1[102] and the protein product was observed to correct cells from individuals showing deficiency of XPG complementation group, which are highly UV sensitive and NER-deficient.[103,104] Studies on XPG protein revealed that it is an 1186 residue containing structure-specific nuclease.[105-107] XPG functions in the late stage of NER by incising the damaged DNA strand in a transcription bubble structure, zero to two nucleotides from the ssDNA-dsDNA junction.[108] Other roles for XPG have also been indicated, which include transcription[109] and the repair of oxidative base damage.[110,111] XPG protein is a member the FEN-1 (Flap EndoNuclease-1) family of structure-specific nucleases (Fig. 5). Mutations that perturb the nuclease function of the protein cause an XP phenotype, while others that truncate the protein cause XP-CS[112-114] (Fig. 5).

The structure of XPG protein is undetermined, but structures of family members are observed to form an α/β 'saddle' structure built from two regions, N and I, which are separated by a flexible loop of approximately 70 amino acids.[115-118] The XPG sequence differs from the family by containing a large spacer region of over 600 amino acids (the Recognition, R-domain) and a C-terminus that is extended by approximately 145 residues (Fig. 5). A seven-residue deletion in the R-domain and a truncation of 260 residues in the C-terminus can give rise to the XP-CS phenotype.[112,114,119] This indicates that both nonenzymatic regions have critical functions in the cell. The large R-domain sequence does not contain clear sequence similarity to any known structural motifs, but it is required for XPG bind to NER bubble substrates.[120-122] XPG bound on bubble substrate interacts with CSB, which stimulates both bubble DNA binding and ATPase activity of CSB and this interaction is

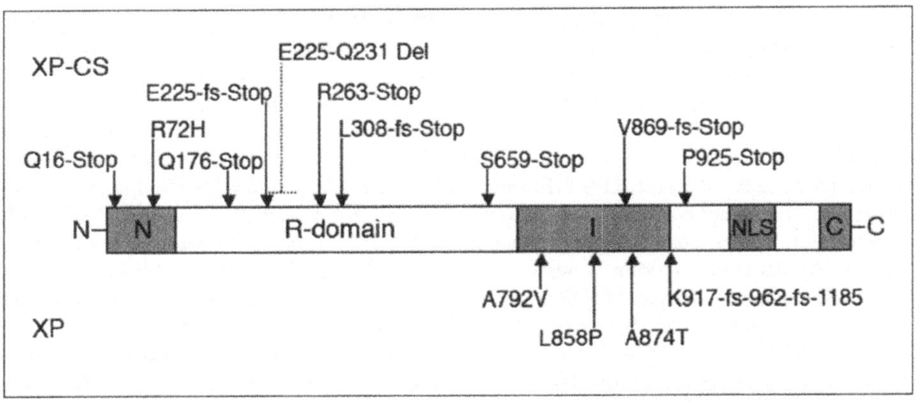

Figure 5. Schematic of XPG endonuclease. The 1186 residue XPG protein is a member of the FEN-1 family of nucleases that contain N and I domains that form the active protein. XPG also has a large spacer region, the R-domain, in addition to a nuclear localization signal, NLS and a C-terminal basic region, labelled C. Mutation in XPG may result in either XP or XP-CS. Examples of XP mutations are shown below the domain schematic; either point mutations or a protein that has two frameshifts (fs) and produces a 1185 amino acid protein with nonXPG amino acids included in the frameshift region cause XP. These XP mutations perturb endonuclease activity, while XP-CS mutations are depicted above the domain schematic and result in a truncated protein.[112-114,119,135,137,138] The single XP-CS point mutation, R72H greatly destabilizes the XPG protein and occurs in a compound heterozygote individual, who also has the Q176-Stop nonsense mutation and XP-CS phenotype.[138] One characterized XP-CS individual produced an infrequent splice variant that deletes residues 225-231 of the R domain and a truncated form of XPG, E225-fs-243-stop.[119]

mediated by the XPG C-terminus extension.[122] This finding implicates XPG function in the early recognition stage of TCR that requires CSB protein. This early stage function is also suggested by interaction of XPG with elongating RNAP II in vivo and with stalled RNAP II in vitro[122] Stalled RNAP II blocks the XPG incision activity in vitro but this can be removed by the addition of TFIIH and ATP.[122] Notably, XPG has also been observed to bind TFIIH.[123] Mutants of XPG protein lacking the C-terminus due to XP-CS mutations did not form a stable complex with TFIIH, while XPG mutant proteins with the classical XP phenotype bind efficiently.[123] Altered XPG protein that failed to associate with TFIIH also caused this dissociation of CAK subunit from the TFIIH core.[123] SiRNA knock down of XPG results in CAK dissociation, while expression of wild-type XPG restores the interactions and TFIIH architecture.[123] Overall, this indicates that the XP-CS phenotype might arise from perturbed interactions with either the CSB protein partner and/or with TFIH in mutant forms of XPG. Moreover, TFIIH is required for transactivation of nuclear receptors and the loss of this activity might cause some of the phenotypes of XP-CS in XPG. This would be similar to XP-CS mutations in XPD, likely altering TFIIH architecture and resulting in the loss of subcutaneous fat tissue and hypogonadism.[71,124,125] However, loss of TFIIH transactivation might not common to all CS-linked mutations, since transactivation is normal in XP-CS of CSA, CSB and XPB mutants. Thus, further studies to define the common transcription/repair deficiencies among the CS-linked proteins are needed.

Perspectives

We have presented the recent findings on the structures and protein partner interactions of the five proteins mutations in which cause either CS or CS syndrome that is combined with XP. These novel studies are providing a considerable increase in our understanding of the molecular basis of the complex disease phenotype of CS. This also includes results defining catalytic mechanisms which allows for comparison with the molecular mechanisms of other key DNA repair proteins, such as RecQ helicases that cause progeria when defective or the Mre11-Rad50-Nbs1 (MRN) complex that can cause neurological disorders when mutated, to define common functionalities and key pathway differences.[126] These types of studies also indicate that the CS-linked proteins and their key complexes are regulating TCR pathway progression. This occurs through protein conformational changes, induced by signaling events that may include ubiquitination and by binding of substrate or protein partners. Thus, continuation of these studies will likely provide a detailed understanding of how mutation of specific gene results in the CS phenotype. This is particularly the case with the XPB, XPD and XPG proteins: the combination of high resolution protein crystallography and lower resolution small angle X-ray scattering studies[127,128] will likely define how CS mutations uncouple enzymatic activities and perturb conformational changes and/or protein partner interactions. However, molecular studies on CSB and probably CSA will also have to include new analyses of genetic and/or environmental factors. This is because the same inactivating mutation in CSB has been observed to affect XP and/or CS phenotypes.[129]

A current model for CS is that neurodegeneration may be caused by an apoptotic response to lesions blocking transcription. The brain, with its limited capacity for cellular proliferation, therefore undergoes a loss of cells leading to the CS phenotype. However, alternate models for the CS phenotypes also warrant further investigation. One group has suggested that the CS phenotype may be due deregulation of additional downstream targets of the CSA E3 ubiquitination system.[130] Supporting this is their observation of increased protein levels in CS cells, including p21 protein that is known to result reduced repair and increased oxidative damage.[130] A second group has reported breaks in the DNA of XPD cells with XP-CS mutation that are not localized at sites of DNA damage;[83,131] instead, they probably occur at sites of transcription initiation and thus may be partially responsible for XP-CS phenotype specifically in XPD.

Interestingly, the increased apoptosis in CS cells might be reason behind the lack of tumorigenesis in CS individual, despite increased rates of point mutations in their DNA. This feature, along with the sensitivity of NER deficient cells to interstrand cross-linkers that are used as therapeutic against cancer, suggests the CS proteins and their partners in DNA repair pathway progression may also be attractive targets for the generation of new anti-tumorigenic compounds. Thus, continued structural based approaches on CS proteins might also be of benefit to potential rational-based drug design efforts for therapeutics against cancer.

Acknowledgements

Studies in the Tainer lab related to this chapter are supported by U.S. National Institute of Health grants CA104660, CA92584, CA112093 and CA081967.. Support is also provided by the US Department of Energy program Integrated Diffraction Analysis Technologies (IDAT) and the DOE program Molecular Assemblies Genes and Genomics Integrated Efficiently (MAGGIE), under Contract Number DE-AC02-05CH11231. Support for these studies is also provided in part by the Human Frontier Science Program.

References

1. Batty DP, Wood RD. Damage recognition in nucleotide excision repair of DNA. Gene 2000; 241(2):193-204.
2. Hanawalt PC. Subpathways of nucleotide excision repair and their regulation. Oncogene 2002; 21(58):8949-8956.
3. Mellon I, Spivak G, Hanawalt PC. Selective removal of transcription-blocking DNA damage from the transcribed strand of the mammalian DHFR gene. Cell 1987; 51(2):241-249.
4. Tornaletti S, Hanawalt PC. Effect of DNA lesions on transcription elongation. Biochimie 1999; 81(1-2):139-146.
5. Svejstrup JQ. Rescue of arrested RNA polymerase II complexes. J Cell Sci 2003; 116(Pt 3):447-451.
6. Ljungman M, Lane DP. Transcription—guarding the genome by sensing DNA damage. Nat Rev Cancer 2004; 4(9):727-737.
7. Lehmann AR. DNA repair-deficient diseases, xeroderma pigmentosum, Cockayne syndrome and trichothiodystrophy. Biochimie 2003; 85(11):1101-1111.
8. Kraemer KH, Patronas NJ, Schiffmann R et al. Xeroderma pigmentosum, trichothiodystrophy and Cockayne syndrome: a complex genotype-phenotype relationship. Neuroscience 2007; 145(4):1388-1396.
9. Mayne LV, Lehmann AR. Failure of RNA synthesis to recover after UV irradiation: an early defect in cells from individuals with Cockayne's syndrome and xeroderma pigmentosum. Cancer Res 1982; 42(4):1473-1478.
10. de Waard H, de Wit J, Andressoo JO et al. Different effects of CSA and CSB deficiency on sensitivity to oxidative DNA damage. Mol Cell Biol 2004; 24(18):7941-7948.
11. Spivak G, Hanawalt PC. Host cell reactivation of plasmids containing oxidative DNA lesions is defective in Cockayne syndrome but normal in UV-sensitive syndrome fibroblasts. DNA Repair (Amst) 2006; 5(1):13-22.
12. D'Errico M, Parlanti E, Teson M et al. The role of CSA in the response to oxidative DNA damage in human cells. Oncogene 2007; 26(30):4336-4343.
13. Henning KA, Li L, Iyer N et al. The Cockayne syndrome group A gene encodes a WD repeat protein that interacts with CSB protein and a subunit of RNA polymerase II TFIIH. Cell 1995; 82(4):555-564.
14. Stefanini M, Fawcett H, Botta E et al. Genetic analysis of twenty-two patients with Cockayne syndrome. Hum Genet 1996; 97(4):418-423.
15. McDaniel LD, Legerski R, Lehmann AR et al. Confirmation of homozygosity for a single nucleotide substitution mutation in a Cockayne syndrome patient using monoallelic mutation analysis in somatic cell hybrids. Hum Mutat 1997; 10(4):317-321.
16. Ren Y, Saijo M, Nakatsu Y et al. Three novel mutations responsible for Cockayne syndrome group A. Genes Genet Syst 2003; 78(1):93-102.
17. Cao H, Williams C, Carter M et al. CKN1 (MIM 216400): mutations in Cockayne syndrome type A and a new common polymorphism. J Hum Genet 2004; 49(1):61-63.
18. Ridley AJ, Colley J, Wynford-Thomas D et al. Characterisation of novel mutations in Cockayne syndrome type A and xeroderma pigmentosum group C subjects. J Hum Genet 2005; 50(3):151-154.
19. Kleppa L, Kanavin OJ, Klungland A et al. A novel splice site mutation in the Cockayne syndrome group A gene in two siblings with Cockayne syndrome. Neuroscience 2007; 145(4):1397-1406.
20. Zhou HX, Wang G. Predicted structures of two proteins involved in human diseases. Cell Biochem Biophys 2001; 35(1):35-47.
21. Li D, Roberts R. WD-repeat proteins: structure characteristics, biological function and their involvement in human diseases. Cell Mol Life Sci 2001; 58(14):2085-2097.
22. Neer EJ, Schmidt CJ, Nambudripad R et al. The ancient regulatory-protein family of WD-repeat proteins. Nature 1994; 371(6495):297-300.

23. Lambright DG, Sondek J, Bohm A et al. The 2.0A crystal structure of a heterotrimeric G protein. Nature 1996; 379(6563):311-319.
24. Sondek J, Bohm A, Lambright DG et al. Crystal structure of a G-protein beta gamma dimer at 2.1A resolution. Nature 1996; 379(6563):369-374.
25. Sprague ER, Redd MJ, Johnson AD et al. Structure of the C-terminal domain of Tup1, a corepressor of transcription in yeast. EMBO J 2000; 19(12):3016-3027.
26. Voegtli WC, Madrona AY, Wilson DK. The structure of Aip1p, a WD repeat protein that regulates Cofilin-mediated actin depolymerization. J Biol Chem 2003; 278(36):34373-34379.
27. Madrona AY, Wilson DK. The structure of Ski8p, a protein regulating mRNA degradation: Implications for WD protein structure. Protein Sci 2004; 13(6):1557-1565.
28. Saijo M, Hirai T, Ogawa A et al. Functional TFIIH is required for UV-induced translocation of CSA to the nuclear matrix. Mol Cell Biol 2007; 27(7):2538-2547.
29. Fousteri M, Vermeulen W, van Zeeland AA et al. Cockayne syndrome A and B proteins differentially regulate recruitment of chromatin remodeling and repair factors to stalled RNA polymerase II in vivo. Mol Cell 2006; 23(4):471-482.
30. Nakatsu Y, Asahina H, Citterio E et al. XAB2, a novel tetratricopeptide repeat protein involved in transcription-coupled DNA repair and transcription. J Biol Chem 2000; 275(45):34931-34937.
31. Groisman R, Polanowska J, Kuraoka I et al. The ubiquitin ligase activity in the DDB2 and CSA complexes is differentially regulated by the COP9 signalosome in response to DNA damage. Cell 2003; 113(3):357-367.
32. Groisman R, Kuraoka I, Chevallier O et al. CSA-dependent degradation of CSB by the ubiquitin-proteasome pathway establishes a link between complementation factors of the Cockayne syndrome. Genes Dev 2006; 20(11):1429-1434.
33. Sugasawa K, Okuda Y, Saijo M et al. UV-induced ubiquitylation of XPC protein mediated by UV-DDB-ubiquitin ligase complex. Cell 2005; 121(3):387-400.
34. Horibata K, Iwamoto Y, Kuraoka I et al. Complete absence of Cockayne syndrome group B gene product gives rise to UV-sensitive syndrome but not Cockayne syndrome. Proc Natl Acad Sci USA 2004; 101(43):15410-15415.
35. van Hoffen A, Venema J, Meschini R et al. Transcription-coupled repair removes both cyclobutane pyrimidine dimers and 6-4 photoproducts with equal efficiency and in a sequential way from transcribed DNA in xeroderma pigmentosum group C fibroblasts. EMBO J 1995; 14(2):360-367.
36. Hashimoto S, Suga T, Kudo E et al. Adult-Onset Neurological Degeneration in a Patient with Cockayne Syndrome and a Null Mutation in the CSB Gene. J Invest Dermatol 2008; 128(6):1597-1599.
37. Licht CL, Stevnsner T, Bohr VA. Cockayne syndrome group B cellular and biochemical functions. Am J Hum Genet 2003; 73(6):1217-1239.
38. Troelstra C, Landsvater RM, Wiegant J et al. Localization of the nucleotide excision repair gene ERCC6 to human chromosome 10q11-q21. Genomics 1992; 12(4):745-749.
39. Troelstra C, van Gool A, de Wit J et al. ERCC6, a member of a subfamily of putative helicases, is involved in Cockayne's syndrome and preferential repair of active genes. Cell 1992; 71(6):939-953.
40. Christiansen M, Thorslund T, Jochimsen B et al. The Cockayne syndrome group B protein is a functional dimer. Febs J 2005; 272(17):4306-4314.
41. Selby CP, Sancar A. Human transcription-repair coupling factor CSB/ERCC6 is a DNA-stimulated ATPase but is not a helicase and does not disrupt the ternary transcription complex of stalled RNA polymerase II. J Biol Chem 1997; 272(3):1885-1890.
42. Citterio E, Rademakers S, van der Horst GT et al. Biochemical and biological characterization of wild-type and ATPase-deficient Cockayne syndrome B repair protein. J Biol Chem 1998; 273(19):11844-11851.
43. Christiansen M, Stevnsner T, Modin C et al. Functional consequences of mutations in the conserved SF2 motifs and posttranslational phosphorylation of the CSB protein. Nucleic Acids Res 2003; 31(3):963-973.
44. Bird LE, Subramanya HS, Wigley DB. Helicases: a unifying structural theme? Curr Opin Struct Biol 1998; 8(1):14-18.
45. Caruthers JM, McKay DB. Helicase structure and mechanism. Curr Opin Struct Biol 2002; 12(1):123-133.
46. Brosh RM Jr, Balajee AS, Selzer RR et al. The ATPase domain but not the acidic region of Cockayne syndrome group B gene product is essential for DNA repair. Mol Biol Cell 1999; 10(11):3583-3594.
47. Balajee AS, Proietti De Santis L, Brosh RM Jr et al. Role of the ATPase domain of the Cockayne syndrome group B protein in UV induced apoptosis. Oncogene 2000; 19(4):477-489.
48. Selzer RR, Nyaga S, Tuo J et al. Differential requirement for the ATPase domain of the Cockayne syndrome group B gene in the processing of UV-induced DNA damage and 8-oxoguanine lesions in human cells. Nucleic Acids Res 2002; 30(3):782-793.

49. Tuo J, Muftuoglu M, Chen C et al. The Cockayne syndrome group B gene product is involved in general genome base excision repair of 8-hydroxyguanine in DNA. J Biol Chem 2001; 276(49):45772-45779.
50. Muftuoglu M, Selzer R, Tuo J et al. Phenotypic consequences of mutations in the conserved motifs of the putative helicase domain of the human Cockayne syndrome group B gene. Gene 2002; 283(1-2):27-40.
51. Tuo J, Chen C, Zeng X et al. Functional crosstalk between hOgg1 and the helicase domain of Cockayne syndrome group B protein. DNA Repair (Amst) 2002; 1(11):913-927.
52. Sunesen M, Selzer RR, Brosh RM Jr et al. Molecular characterization of an acidic region deletion mutant of Cockayne syndrome group B protein. Nucleic Acids Res 2000; 28(16):3151-3159.
53. Muftuoglu M, Sharma S, Thorslund T et al. Cockayne syndrome group B protein has novel strand annealing and exchange activities. Nucleic Acids Res 2006; 34(1):295-304.
54. Citterio E, Van Den Boom V, Schnitzler G et al. ATP-dependent chromatin remodeling by the Cockayne syndrome B DNA repair-transcription-coupling factor. Mol Cell Biol 2000; 20(20):7643-7653.
55. Fousteri M, van Hoffen A, Vargova H et al. Repair of DNA lesions in chromosomal DNA impact of chromatin structure and Cockayne syndrome proteins. DNA Repair (Amst) 2005; 4(8):919-925.
56. Tantin D, Kansal A, Carey M. Recruitment of the putative transcription-repair coupling factor CSB/ERCC6 to RNA polymerase II elongation complexes. Mol Cell Biol 1997; 17(12):6803-6814.
57. Bradsher J, Auriol J, Proietti de Santis L et al. CSB is a component of RNA pol I transcription. Mol Cell 2002; 10(4):819-829.
58. Balajee AS, May A, Dianov GL et al. Reduced RNA polymerase II transcription in intact and permeabilized Cockayne syndrome group B cells. Proc Natl Acad Sci USA 1997; 94(9):4306-4311.
59. Tantin D. RNA polymerase II elongation complexes containing the Cockayne syndrome group B protein interact with a molecular complex containing the transcription factor IIH components xeroderma pigmentosum B and p62. J Biol Chem 1998; 273(43):27794-27799.
60. van den Boom V, Citterio E, Hoogstraten D et al. DNA damage stabilizes interaction of CSB with the transcription elongation machinery. J Cell Biol 2004; 166(1):27-36.
61. Laine JP, Egly JM. Initiation of DNA repair mediated by a stalled RNA polymerase IIO. EMBO J 2006; 25(2):387-397.
62. Flohr C, Burkle A, Radicella JP et al. Poly(ADP-ribosyl)ation accelerates DNA repair in a pathway dependent on Cockayne syndrome B protein. Nucleic Acids Res 2003; 31(18):5332-5337.
63. Thorslund T, von Kobbe C, Harrigan JA et al. Cooperation of the Cockayne syndrome group B protein and poly(ADP-ribose) polymerase 1 in the response to oxidative stress. Mol Cell Biol 2005; 25(17):7625-7636.
64. Wong HK, Muftuoglu M, Beck G et al. Cockayne syndrome B protein stimulates apurinic endonuclease 1 activity and protects against agents that introduce base excision repair intermediates. Nucleic Acids Res 2007; 35(12):4103-4113.
65. Weeda G, Eveno E, Donker I et al. A mutation in the XPB/ERCC3 DNA repair transcription gene, associated with trichothiodystrophy. Am J Hum Genet 1997; 60(2):320-329.
66. Schaeffer L, Roy R, Humbert S et al. DNA repair helicase: a component of BTF2 (TFIIH) basic transcription factor. Science 1993; 260(5104):58-63.
67. Schultz P, Fribourg S, Poterszman A et al. Molecular structure of human TFIIH. Cell 2000; 102(5):599-607.
68. Chang WH, Kornberg RD. Electron crystal structure of the transcription factor and DNA repair complex, core TFIIH. Cell 2000; 102(5):609-613.
69. Araujo SJ, Tirode F, Coin F et al. Nucleotide excision repair of DNA with recombinant human proteins: definition of the minimal set of factors, active forms of TFIIH and modulation by CAK. Genes Dev 2000; 14(3):349-359.
70. Coin F, Proietti De Santis L, Nardo T et al. p8/TTD-A as a repair-specific TFIIH subunit. Mol Cell 2006; 21(2):215-226.
71. Keriel A, Stary A, Sarasin A et al. XPD mutations prevent TFIIH-dependent transactivation by nuclear receptors and phosphorylation of RARalpha. Cell 2002; 109(1):125-135.
72. Dubaele S, Proietti De Santis L, Bienstock RJ et al. Basal transcription defect discriminates between xeroderma pigmentosum and trichothiodystrophy in XPD patients. Mol Cell 2003; 11(6):1635-1646.
73. Weeda G, Wiegant J, van der Ploeg M et al. Localization of the xeroderma pigmentosum group B-correcting gene ERCC3 to human chromosome 2q21. Genomics 1991; 10(4):1035-1040.
74. Guzder SN, Sung P, Bailly V et al. RAD25 is a DNA helicase required for DNA repair and RNA polymerase II transcription. Nature 1994; 369(6481):578-581.
75. Hwang JR, Moncollin V, Vermeulen W et al. A 3′ → 5′ XPB helicase defect in repair/transcription factor TFIIH of xeroderma pigmentosum group B affects both DNA repair and transcription. J Biol Chem 1996; 271(27):15898-15904.

76. Tirode F, Busso D, Coin F et al. Reconstitution of the transcription factor TFIIH: assignment of functions for the three enzymatic subunits, XPB, XPD and cdk7. Mol Cell 1999; 3(1):87-95.
77. Jawhari A, Laine JP, Dubaele S et al. p52 mediates XPB function within the transcription/repair factor TFIIH. J Biol Chem 2002; 277(35):31761-31767.
78. Coin F, Auriol J, Tapias A et al. Phosphorylation of XPB helicase regulates TFIIH nucleotide excision repair activity. EMBO J 2004; 23(24):4835-4846.
79. Fan L, Arvai AS, Cooper PK et al. Conserved XPB core structure and motifs for DNA unwinding: implications for pathway selection of transcription or excision repair. Mol Cell 2006; 22(1):27-37.
80. Obmolova G, Ban C, Hsieh P et al. Crystal structures of mismatch repair protein MutS and its complex with a substrate DNA. Nature 2000; 407(6805):703-710.
81. Soultanas P, Wigley DB. DNA helicases: 'inching forward'. Curr Opin Struct Biol 2000; 10(1):124-128.
82. Soultanas P, Wigley DB. Unwinding the 'Gordian knot' of helicase action. Trends Biochem Sci 2001; 26(1):47-54.
83. Theron T, Fousteri MI, Volker M et al. Transcription-associated breaks in xeroderma pigmentosum group D-cells from patients with combined features of xeroderma pigmentosum and Cockayne syndrome. Mol Cell Biol 2005; 25(18):8368-8378.
84. Singleton MR, Dillingham MS, Wigley DB. Structure and mechanism of helicases and nucleic acid translocases. Annu Rev Biochem 2007; 76:23-50.
85. Kim TK, Ebright RH, Reinberg D. Mechanism of ATP-dependent promoter melting by transcription factor IIH. Science 2000; 288(5470):1418-1422.
86. Sung P, Bailly V, Weber C et al. Human xeroderma pigmentosum group D gene encodes a DNA helicase. Nature 1993; 365(6449):852-855.
87. Weber CA, Salazar EP, Stewart SA et al. ERCC2: cDNA cloning and molecular characterization of a human nucleotide excision repair gene with high homology to yeast RAD3. EMBO J 1990; 9(5):1437-1447.
88. Reardon JT, Ge H, Gibbs E et al. Isolation and characterization of two human transcription factor IIH (TFIIH)-related complexes: ERCC2/CAK and TFIIH. Proc Natl Acad Sci USA 1996; 93(13):6482-6487.
89. Coin F, Marinoni JC, Rodolfo C et al. Mutations in the XPD helicase gene result in XP and TTD phenotypes, preventing interaction between XPD and the p44 subunit of TFIIH. Nat Genet 1998; 20(2):184-188.
90. Schaeffer L, Moncollin V, Roy R et al. The ERCC2/DNA repair protein is associated with the class II BTF2/TFIIH transcription factor. EMBO J 1994; 13(10):2388-2392.
91. Rudolf J, Makrantoni V, Ingledew WJ et al. The DNA repair helicases XPD and FancJ have essential iron-sulfur domains. Mol Cell 2006; 23(6):801-808.
92. Coin F, Marinoni JC, Egly JM. Mutations in XPD helicase prevent its interaction and regulation by p44, another subunit of TFIIH, resulting in Xeroderma pigmentosum (XP) and trichothiodystrophy (TTD) phenotypes. Pathol Biol (Paris) 1998; 46(9):679-680.
93. Imlay JA. Iron-sulphur clusters and the problem with oxygen. Mol Microbiol 2006; 59(4):1073-82.
94. Boal AK, Yavin E, Lukianova OA et al. DNA-bound redox activity of DNA repair glycosylases containing [4Fe-4S] clusters. Biochemistry 2005; 44(23):8397-8407.
95. Kuo CF, McRee DE, Fisher CL et al. Atomic structure of the DNA repair [4Fe-4S] enzyme endonuclease III. Science 1992; 258(5081):434-440.
96. Lehmann AR. The xeroderma pigmentosum group D (XPD) gene: one gene, two functions, three diseases. Genes Dev 2001; 15(1):15-23.
97. Botta E, Nardo T, Lehmann AR et al. Reduced level of the repair/transcription factor TFIIH in trichothiodystrophy. Hum Mol Genet 2002; 11(23):2919-2928.
98. Pugh RA, Honda M, Leesley H et al. The iron-containing domain is essential in Rad3 helicases for coupling of ATP hydrolysis to DNA translocation and for targeting the helicase to the single-stranded DNA-double-stranded DNA junction. J Biol Chem 2008; 283(3):1732-1743.
99. Bootsma D, Hoeijmakers JH. DNA repair. Engagement with transcription. Nature 1993; 363(6425):114-115.
100. Sung P, Higgins D, Prakash L et al. Mutation of lysine-48 to arginine in the yeast RAD3 protein abolishes its ATPase and DNA helicase activities but not the ability to bind ATP. EMBO J 1988; 7(10):3263-3269.
101. Winkler GS, Araujo SJ, Fiedler U et al. TFIIH with inactive XPD helicase functions in transcription initiation but is defective in DNA repair. J Biol Chem 2000; 275(6):4258-4266.
102. Takahashi E, Shiomi N, Shiomi T. Precise localization of the excision repair gene, ERCC5, to human chromosome 13q32.3-q33.1 by direct R-banding fluorescence in situ hybridization. Jpn J Cancer Res 1992; 83(11):1117-1119.

103. Scherly D, Nouspikel T, Corlet J et al. Complementation of the DNA repair defect in xeroderma pigmentosum group G cells by a human cDNA related to yeast RAD2 Nature 1993; 363(6425):182-185.
104. O'Donovan A, Wood RD. Identical defects in DNA repair in xeroderma pigmentosum group G and rodent ERCC group 5. Nature 1993; 363(6425):185-188.
105. Shiomi T, Harada Y, Saito T et al. An ERCC5 gene with homology to yeast RAD2 is involved in group G xeroderma pigmentosum. Mutat Res 1994; 314(2):167-175.
106. Habraken Y, Sung P, Prakash L et al. Human xeroderma pigmentosum group G gene encodes a DNA endonuclease. Nucleic Acids Res 1994; 22(16):3312-3316.
107. Cloud KG, Shen B, Strniste GF et al. XPG protein has a structure-specific endonuclease activity. Mutat Res 1995; 347(2):55-60.
108. O'Donovan A, Davies AA, Moggs JG et al. XPG endonuclease makes the 3' incision in human DNA nucleotide excision repair. Nature 1994; 371(6496):432-435.
109. Lee SK, Yu SL, Prakash L et al. Requirement of yeast RAD2, a homolog of human XPG gene, for efficient RNA polymerase II transcription. implications for Cockayne syndrome. Cell 2002; 109(7):823-834.
110. Klungland A, Hoss M, Gunz D et al. Base excision repair of oxidative DNA damage activated by XPG protein. Mol Cell 1999; 3(1):33-42.
111. Bessho T. Nucleotide excision repair 3' endonuclease XPG stimulates the activity of base excision repairenzyme thymine glycol DNA glycosylase. Nucleic Acids Res 1999; 27(4):979-983.
112. Emmert S, Slor H, Busch DB et al. Relationship of neurologic degeneration to genotype in three xeroderma pigmentosum group G patients. J Invest Dermatol 2002; 118(6):972-982.
113. Lalle P, Nouspikel T, Constantinou A et al. The founding members of xeroderma pigmentosum group G produce XPG protein with severely impaired endonuclease activity. J Invest Dermatol 2002; 118(2):344-351.
114. Nouspikel T, Lalle P, Leadon SA et al. A common mutational pattern in Cockayne syndrome patients from xeroderma pigmentosum group G: implications for a second XPG function. Proc Natl Acad Sci USA 1997; 94(7):3116-3121.
115. Ceska TA, Sayers JR, Stier G et al. A helical arch allowing single-stranded DNA to thread through T5 5'-exonuclease. Nature 1996; 382(6586):90-93.
116. Hosfield DJ, Mol CD, Shen B et al. Structure of the DNA repair and replication endonuclease and exonuclease FEN-1: coupling DNA and PCNA binding to FEN-1 activity. Cell 1998; 95(1):135-146.
117. Mueser TC, Nossal NG, Hyde CC. Structure of bacteriophage T4 RNase H, a 5' to 3' RNA-DNA and DNA-DNA exonuclease with sequence similarity to the RAD2 family of eukaryotic proteins. Cell 1996; 85(7):1101-1112.
118. Hwang KY, Baek K, Kim HY et al. The crystal structure of flap endonuclease-1 from Methanococcus jannaschii. Nat Struct Biol 1998; 5(8):707-713.
119. Thorel F, Constantinou A, Dunand-Sauthier I et al. Definition of a short region of XPG necessary for TFIIH interaction and stable recruitment to sites of UV damage. Mol Cell Biol 2004; 24(24):10670-10680.
120. Dunand-Sauthier I, Hohl M, Thorel F et al. The spacer region of XPG mediates recruitment to nucleotide excision repair complexes and determines substrate specificity. J Biol Chem 2005; 280(8):7030-7037.
121. Hohl M, Dunand-Sauthier I, Staresincic L et al. Domain swapping between FEN-1 and XPG defines regions in XPG that mediate nucleotide excision repair activity and substrate specificity. Nucleic Acids Res 2007; 35(9):3053-3063.
122. Sarker AH, Tsutakawa SE, Kostek S et al. Recognition of RNA polymerase II and transcription bubbles by XPG, CSB and TFIIH: insights for transcription-coupled repair and Cockayne syndrome. Mol Cell 2005; 20(2):187-198.
123. Ito S, Kuraoka I, Chymkowitch P et al. XPG stabilizes TFIIH, allowing transactivation of nuclear receptors: implications for Cockayne syndrome in XP-G/CS patients. Mol Cell 2007; 26(2):231-243.
124. Compe E, Drane P, Laurent C et al. Dysregulation of the peroxisome proliferator-activated receptor target genes by XPD mutations. Mol Cell Biol 2005; 25(14):6065-6076.
125. Drane P, Compe E, Catez P et al. Selective regulation of vitamin D receptor-responsive genes by TFIIH. Mol Cell 2004; 16(2):187-197.
126. Perry JJ, Fan L, Tainer JA. Developing master keys to brain pathology, cancer and aging from the structural biology of proteins controlling reactive oxygen species and DNA repair. Neuroscience 2007; 145(4):1280-1299.
127. Tsutakawa SE, Hura GL, Frankel KA et al. Structural analysis of flexible proteins in solution by small angle X-ray scattering combined with crystallography. J Struct Biol 2007; 158(2):214-223.
128. Putnam CDH M, Hura GL, Tainer JA. X-ray solution scattering (SAXS) combined with crystallography and computation: defining accurate macromolecular structures, conformations and assemblies in solution. Q Rev Biophys 2008; 40(3):191-295.

129. Colella S, Nardo T, Botta E et al. Identical mutations in the CSB gene associated with either Cockayne syndrome or the DeSanctis-cacchione variant of xeroderma pigmentosum. Hum Mol Genet 2000; 9(8):1171-1175.
130. Cleaver JE, Hefner E, Laposa RR et al. Cockayne syndrome exhibits dysregulation of p21 and other gene products that may be independent of transcription-coupled repair. Neuroscience 2007; 145(4):1300-1308.
131. Berneburg M, Lowe JE, Nardo T et al. UV damage causes uncontrolled DNA breakage in cells from patients with combined features of XP-D and Cockayne syndrome. EMBO J 2000; 19(5):1157-1166.
132. Colella S, Nardo T, Mallery D et al. Alterations in the CSB gene in three Italian patients with the severe form of Cockayne syndrome (CS) but without clinical photosensitivity. Hum Mol Genet 1999; 8(5):935-941.
133. Mallery DL, Tanganelli B, Colella S et al. Molecular analysis of mutations in the CSB (ERCC6) gene in patients with Cockayne syndrome. Am J Hum Genet 1998; 62(1):77-85.
134. Broughton BC, Steingrimsdottir H, Weber CA et al. Mutations in the xeroderma pigmentosum group D DNA repair/transcription gene in patients with trichothiodystrophy. Nat Genet 1994; 7(2):189-194.
135. Cleaver JE, Thompson LH, Richardson AS et al. A summary of mutations in the UV-sensitive disorders: xeroderma pigmentosum, Cockayne syndrome and trichothiodystrophy. Hum Mutat 1999; 14(1):9-22.
136. Broughton BC, Berneburg M, Fawcett H et al. Two individuals with features of both xeroderma pigmentosum and trichothiodystrophy highlight the complexity of the clinical outcomes of mutations in the XPD gene. Hum Mol Genet 2001; 10(22):2539-2547.
137. Nouspikel T, Clarkson SG. Mutations that disable the DNA repair gene XPG in a xeroderma pigmentosum group G patient. Hum Mol Genet 1994; 3(6):963-967.
138. Zafeiriou DI, Thorel F, Andreou A et al. Xeroderma pigmentosum group G with severe neurological involvement and features of Cockayne syndrome in infancy. Pediatr Res 2001; 49(3):407-412.

Chapter 7

Cockayne Syndrome: Its Overlap with Xeroderma Pigmentosum and Other Progeroid Syndromes

W. Clark Lambert,* Edward G. Neilan, Claude E. Gagna, George G. Kihiczak, Hon-Reen Kuo and Muriel W. Lambert

Introduction

Although this chapter is devoted to Cockayne syndrome (CS), there are, in fact, at least three disorders, CS, trichothiodystrophy (TTD) and xeroderma pigmentosum (XP), the etiopathogenesis of which are involved with the same biochemical pathways and, in a number of cases, with the same gene(s).[1-5] In some instances, patients have unequivocal evidence of having more than one of these diseases, which are clinically quite heterogeneous. This chapter will review some of these unusual cases, for CS known as XP/CS overlap syndromes or XP/CS overlap complexes.

In theory, there are at least two ways an individual could be affected by these very different diseases (i.e., CS and XP). The first is that he or she could simply be extraordinarily unlucky and be struck by two genetic bullets. This idea implies that the individual is homozygous or hemizygous (in the case that an allele on a sex chromosome is involved) for defective alleles at two different loci. This occurrence, which two of us (WCL and MWL) have previously labeled "corecessive inheritance", would imply extremely high carrier frequencies for at least some of these defective alleles.[6,7] However, there is no direct evidence that corecessive inheritance is responsible for these overlap syndromes, although there are a number of examples in which two individuals with identical mutations have very different clinical phenotypes, implying involvement of more than one gene.

Another way one could be affected by more than one of these disorders is that one is homozygous or hemizygous for defective alleles at a single locus, but the gene product associated with that locus is involved in more than one biochemical pathway within the cell. There is considerable evidence that this is the etiopathogenesis of at least some of the overlap syndromes, which, although quite rare, are much more common than would be expected, based on the frequencies of CS and XP considered independently.

In particular, we will be concerned with two pathways, initiation, and to some extent continuation, of transcription, especially transcription by RNA polymerase II, which transcribes mRNA for active genes within the cell, and nucleotide excision repair (NER), which is defective in all complementation groups of XP except the variant group XPV. As has been reviewed extensively elsewhere in this volume, the large RNA transcription subunit, TFIIH, is intimately involved in both pathways, so much so that it could just as easily be considered a DNA repair protein complex as an RNA transcription protein complex. It is also involved in other cellular processes.

TFIIH is a complex containing at least 10 proteins, notably including XPB, XPD and XPG, mutations in which appears to account for almost all of these overlap syndromes. It is also bound indirectly to both the ERCC6 (CSB) and the ERCC8 (CSA) gene products. The consensus, if

*Corresponding Author: W. Clark Lambert—Department of Pathology and Laboratory Medicine, UMDNJ-New Jersey Medical School. Email: lamberwc@umdnj.edu

Molecular Mechanisms of Cockayne Syndrome, edited by Shamim I. Ahmad.

there is such a thing, is that mutations in these genes that cause XP are those that affect the NER mechanism (particularly global genomic NER (GG-NER)). On the other hand, those mutations that primarily affect transcription-coupled NER (TC-NER), as well as transcription and possibly other systems cause TTD or CS. Presumably the overlap syndromes are due to mutations that affect more than one pathway and we shall regard this as a working hypothesis.

However, for several reasons it is quite likely the above hypothesis is an oversimplification: first, there are many other biochemical pathways that are either known to include or are likely to be found to involve these same gene products. These include both short patch and long patch base excision repair (BER) of adducts in DNA, mismatch repair (MMR) of mismatched bases, homologous recombination (HR) and nonhomologous end-joining (NHEJ) of DNA double strand breaks (DSBs). Other processes that interact with these include telomere maintenance and cell cycle control mechanisms. Second, many nuclear proteins are, at least to some extent, interdependent for maintenance of stability and other functions, so that a protein may affect one or more of these pathways without actually being involved in it directly. A well known example is the Fanconi anemia proteins, deficiency of one of which leads to accelerated breakdown of others in many instances. Third, nuclear proteins/gene products may affect rates of transcription or translation of other genes or rates of activation of their gene products. Fourth, the function of some of these proteins, or even the entire pathway in which they are functioning, remains unclear. The list goes on and on. We have shown, for example, that a major nuclear structural protein, α-spectrin II, is active in the Fanconi anemia pathway, the function of which is still poorly understood.[8]

There is yet another area of uncertainty that must be addressed before we attempt to review these overlap syndromes: the diagnosis of any of these is dependent on clinical assessment, on who is looking and how hard they are doing so. Once this has been done, there is also a question regarding classification, of which signs and symptoms rate classification where and how. Is one adult patient with early freckling but no skin cancers comparable to another with severe dyspigmentation and dozens of cancers arising in childhood? Should one be considered as having XP and another not? In these classifications both are likely to be considered to have XP, with additional signs/symptoms, which are similarly variable, causing them to be classified as also having CS or TTD. One must review these reports, many of which are authored by biochemists and molecular biologists who are wonderfully competent in their fields but who are not necessarily similarly skilled clinicians, or who may not even have direct access to the relevant clinical information, with caution. There is also heterogeneity in competencies among physicians reporting these cases. For example, a neurologist is less likely to note freckling than a dermatologist who, in turn, is less likely to detect mild spasticity.

Although there are reports of XP/TTD and XP/CS overlap syndromes, we are not aware of a report of a "TTD/CS overlap syndrome" or of an "XP/TTD/CS overlap syndrome". However, this may have been overlooked because the authors of the reports were mostly focusing on XP/TTD overlap or XP/CS overlap rather than the other possibilities and may therefore represent a pedagogic solution to a scientific problem. Clinically, there is considerable overlap between TTD and CS and both are presumably (see "working hypothesis" above) related to transcriptional deficiencies. XP/TTD overlap syndromes have been reviewed elsewhere.[9]

In the following we shall review not only XP/CS overlap syndromes due to XPB, XPD and XPG mutations, but certain other overlaps as well. In particular, we shall examine overlaps of CS and XP with progeroid syndromes. CS and TTD may themselves be considered progeroid syndromes and these syndromes, although imperfectly so, are nevertheless intriguing models for ageing. We shall also review the ultraviolet sensitivity syndrome (UVSS), which has been reported to be caused by a total lack of the CSB protein.

The Ultraviolet-Sensitive Syndrome (UVSS)

Fujiwara et al described a patient with clinical manifestations similar to XP, including acute sunburn following minimal sun exposure, freckling, xerosis (dryness) telangiectasias and dyspigmentation of sun-exposed skin but without development of tumors or neurological disorders.[10] Subsequently, Itoh et al described additional patients.[11] A delayed recovery of RNA synthesis, following UV exposure, was noted in the patients' cells—a feature of CS. These authors proposed the new term "ultraviolet-sensitive syndrome (UVSS)" to denote these patients.[11,12]

Horibata et al found that cells from the original patient described by Fujiwara et al[10] were homozygous for a null mutation in *ERCC6/CSB* (the excision repair, complementation group 6/ CS, complementation group B gene).[13] A cell line from a different patient with the UVSS had no mutations in *ERCC6/CSB* and normal amounts of the ERCC6/CSB protein, however, indicating a genetic heterogeneity within the UVSS. The UVSS has not yet been reported to have any overlapping signs or symptoms of TTD or CS.

Trichothiodystrophy

The term, trichothiodystrophy (TTD), was introduced by Price et al to describe a syndrome or syndromes characterized by the presence of brittle, sulfur-deficient hair.[14] The hair and nails of the patients suffering from this disease, are brittle because they contain reduced levels of cysteine-rich matrix proteins. Affected individuals also have ichthyosis, which is usually apparent at birth and much less apparent afterwards, and mentally and physically retarded development. TTD is part of a more broadly defined entity known as the IBIDS (ichthyosis, brittle hair, impaired intelligence, decreased fertility and short stature) syndrome,[15] which itself is somewhat poorly defined due to clinical variation between patients.[16] The original case was described by Tay and the disorder is known as the Tay syndrome.[17] The hairs in patients with TTD have a distinctive, diagnostically useful appearance on polarized light microscopy known as the "tiger-tail" anomaly; the disease may also be diagnosed by amino acid analysis of the patients' hair shafts, which show decreased sulphur and cysteine content.[14]

About half of the patients with TTD have also photosensitivity (clinically manifested as sun-sensitivity) and an associated cellular defect in the TCR subtype of NER.[18] However, like Cockayne syndrome, to date TTD has not been reported to develop skin cancer.[16,18] The photosensitive form of TTD is also known as "trichothiodystrophy-photosensitive (TTDP)".[19]

The TTD phenotype can be caused by mutations in at least three different genes. The overwhelming majority of patients have mutations in *ERCC2/XPD* (the Excision Repair Cross-Complementing 2/Xeroderma Pigmentosum Complementation Group D gene). A single family has had, instead, biallelic mutations in the *ERCC3/XPB* (the Excision Repair Cross-Complementing 3/Xeroderma Pigmentosum Complementation Group B gene, see below). These two genes encode the two helicase subunits of the transcription/DNA repair factor, TFIIH (Transcription Factor for RNA polymerase II, subunit H, which itself is composed of at least 10 subunit proteins). A different TTD complementation group was identified by Stefanini et al, when they showed that cells from patients in this new group could complement all XP complementation groups and that this complementation was not intragenic.[19] Cells from this complementation group were originally identified as "TTD1BR". It was suggested that it may be identified as "TTD2", with the other known TTD cases at the time (all due at that time to mutations in *ERCC3/XPD*) identified as in complementation group "TTD1". Following the revision in nomenclature proposed by Lehmann, et al, this group is now known as TTD-A.[20] However, these alternative nomenclatures may be seen in the literature. This complementation group of TTD (TTT-A) has been found to be caused by a mutation in the gene (*GTF2H5/TTD-A*) encoding a newly identified, tenth subunit of TFIIH, TFB5.[21] The overall levels of TFIIH in TTDA cells are also significantly depressed, presumably because TFB5, in addition to any other functions it may have, acts as a stabilization factor for TFIIH.[22] It is not known whether the depressed level of TFIIH or the defect in TFB5, or a combination of these, is responsible for the disease.

Depending upon the mutation and possibly other components of the genetic background in individual cases, mutations in the *XP-B* or *XP-D* may cause XP only, TTD only, or, for *XP-D*, an overlap syndrome incorporating elements of both diseases, known as the xeroderma pigmentosum/ trichothiodystrophy overlap syndrome (XP/TTD syndrome). Mutations in these genes and also in *XP-G* (see below) may also cause CS (notably with the absence of certain features) or a crossover syndrome between XP and CS (See below).

XP/Cockayne (XP/CS) Overlap Syndrome

CS is a rare (approximately 1/100,000 live births) autosomal recessive disorder characterized in part by progressive neurodegeneration, a progerioid (premature aged, wizened) appearance and photosensitivity without development of skin cancer.[23,24] The neurodegeneration is associated with demyelination, rather than with neural cell degeneration, as is seen in cases of XP associated with neurological disease.[25] Three forms of CS have been delineated: CS Type I, also known as "classic CS", CS Type II, "Cerebro-Oculo-Facial-Skeletal syndrome (COFS)", the "Pena-Shokeir syndrome, Type II" or "connatal CS", and CS Type III. Some authors distinguish between CS Type II and COFS; we shall consider them synonymous here. CS Type II is the most severe, with abnormalities manifested in utero and gross deformities present at birth. In CS Type I, abnormalities are not apparent at birth but appear in early childhood. In CS Type III the disease is milder with later onset, in some cases as late as the third decade of life. The XP/CS overlap syndrome may constitute a fourth form, but individual cases may fall within one of the three other forms, particularly the COFS syndrome. When it occurs without sun-sensitivity, CS has variable phenotype.[26]

CS Type I is characterized by normal prenatal growth but with growth and developmental abnormalities arising in the first two years after birth. Following onset, growth retardation becomes severe, so that at the end of the course of the disease, height, weight and head circumference are well below the fifth percentile. A very characteristic finding is recession of the eyes into their sockets, due to lack of post-orbital fat. This, along with lack of fat in facial tissues, contributes to a "wizened" faces forming part of a progeroid phenotype. Dental caries, due to decreased salivation and defective innervation of the salivary glands, is a frequent feature. Patients may have a characteristic "horse rider" gait. Neurodegeneration is progressive and death usually occurs by seven years of age.

COFS (CS Type II) is even more severe than CS Type I, with gross abnormalities becoming apparent in utero. These abnormalities are clearly manifest at birth and progress rapidly afterwards. CS Type III is typically much less severe with signs and symptoms occurring much later in life.

CS of any type tends to be associated with spasticity, increased deep tendon reflexes and joint disease. X-ray examination reveals calcifications of basal ganglia of the brain. Neuropathologic changes are associated with demyelination, rather than the neuronal degeneration characteristic of the neurodegenerative disease seen in some cases of XP. Approximately 75 percent of cases of CS have been found to be due to a biallelic mutation in the *ERCC6/CSB* gene. Most of the remaining 25 percent of cases are due to a mutation in the *ERCC8/CSA* gene. Mutations in *ERCC8/CSA* or *ERCC6/CSB* cause defective TCR of UV damage.[27] There may also be subtle transcription defects.[29-30]

Very few cases of CS are due to a mutation in the *XPB*, *XPD* or *XPG* genes; some of these also have features of XP and are thus examples of the XP/CS overlap syndrome. A tiny subset of cases is due to a gene(s) that has not yet been identified.

Patients with CS, not in complementation groups A or B, were formerly identified as in complementation group C, CS-C.[31] This group has subsequently been abandoned.[20] Some of these patients had biallelic mutations in *XPD*.

Patients with the XP/CS overlap syndrome(s) have very variable phenotypes and it is perhaps unwise to attempt to generalize regarding them. However, their XP features at least include early freckling, accompanied by dryness of the skin in sun-exposed areas and in some cases by skin cancers. Their CS features tend to include mental retardation, spasticity, short stature and hypogonadism, but not skeletal dysplasia, although there are exceptions.

XP-B/CS Overlap Syndrome

To date, seven mutations in *ERCC3/XPB* have been identified in six families.[32] Each family has one or more individuals with either XP (one family), TTD (one family), or the XP/CS overlap syndrome (four families). The very limited range of these mutations may be related to the fact that the XPB protein is a helicase, active in both NER and transcription and essential for both of them. In contrast, the XPD protein, although a helicase active in both NER and transcription, is essential only for the former.[32,33] Thus a mutation in *XPB* is much more likely to be lethal than one in *XPD*.

The six families with XPB mutations have been extensively studied by Oh, et al.[32] The first XPB patient identified,[34,35] (their Family III[32]) was found to have a severe form of the XP/CS overlap syndrome.[36] XP signs/symptoms included extreme sensitivity to sunlight, with blistering noted in infancy, dyspigmentation and multiple skin cancers with onset at age 15 years. CS signs/symptoms included severe dwarfism, microcephaly, microphthalmia, sensorineural deafness, severe mental retardation (IQ of 40 but with good social skills), corneal scarring, cataracts, pigmentary retinal degeneration, optic atrophy, abnormal (flat) retinogram, hypereflexia, ataxia, decreased nerve conduction velocity, enlarged cerebral ventricles (brain), basal ganglia calcifications and immature sexual development. The patient died at 33 years of age of cardiovascular disease.

A second family (Family II[32]) had two brothers with mild XP/CS overlap syndrome. They were originally reported by Scott, et al,[37] with the XPB classification subsequently carried out by Vermeulen et al;[38] The XP features in these patients consisted of severe sunburn at approximately six weeks of age and early freckling-like dyspigmentation. Neither had developed skin cancer when studied at ages 38 and 41, respectively. However, CS features were somewhat more severe, consisting of short stature, sensorineural hearing loss which began at the age of four years and late onset hyperreflexia, other clinical neurological deficiencies, decreased nerve conduction velocity, enlarged central ventricles, demyelinating neuropathy, retinopathology and immature sexual development.

A third family (Family IV[32]) consisted of a 27 year-old man with a severe form of XP/CS overlap syndrome from Slovenia.[39] XP features included sun hypersensitivity, dyspigmentation and atrophy of sun-exposed skin, telangiectasias and recurrent eyelid squamous cell carcinoma. CS features included growth and mental retardation, sensorineural deafness, optic atrophy, pigmentary retinopathy, ataxia, decreased nerve conduction velocity, enlargement of central ventricles, cerebellar atrophy and calcifications of the basal ganglia. He died of end-stage renal failure at 31 years of age.

A fourth family (Family V[32]), consisted of a 10 year-old girl with severe XP/CS overlap syndrome from Germany. XP signs/symptoms consisted of severe sunburn at age two weeks and freckling on sun-exposed skin since early childhood. She had not developed any skin cancers when evaluated at 10 years of age. CS signs/symptoms consisted of increasing growth retardation, bird-like faces, progressive hearing loss beginning at age 7 years and a progressive loss of mental development. Although she could walk at age 20 months, she subsequently developed ataxia, abnormal balance, central coordination disability and intermittent tremor. She had hyperopia but not optic nerve atrophy or cataracts. Patellar and Achilles tendon reflexes were absent; the Babinski sign was negative.

A fifth (Family I) and sixth (Family VI) families reported by Oh, et al[32] had only XP and TTD respectively.

All of these XPB families' cells were hypersensitive to UV irradiation and showed very low unscheduled DNA synthesis (UDS) levels (reduced to 4-10% of normal levels) following UV irradiation. Recovery of RNA synthesis (RRS) was also low following UV irradiation. Oh, et al also showed that more severely affected patients with XP-B mutations had less residual XP-B activity than less severely affected ones.[32]

XP-D/CS Overlap Syndrome

There are three reports of the XP/CS overlap syndrome falling within the D complementation group of XP.[40-42]

Cells derived from Individuals who have clinical features of CS and also skin changes, including skin cancers, typical of XP, have been found to have biochemical features typical of XP and to fall within the D complementation group of XP.[43] Cellular functional assays of cells from some patients with the COFS phenotype show sensitivity to UV, decreased UDS following UV radiation and complementation of these features by the *XP-D* product.[44]

In cell survival assays, cells from XP-D/CS patients show markedly increased hypersensitivity to UV radiation compared to XP cells within the same complementation group.[40,45] This is associated with uncontrolled DNA breakage, unrelated to sites of UV radiation induced pyrimidine dimers or pyrimidine-pyrimidone 6-4 photo-products, found specifically in cells derived from XP/CS overlap syndrome patients. As a rule, the sites of mutations in the *XPD* gene are specific for one or other of the clinical phenotypes.

Whereas two patients with the XP/CS overlap syndrome have had mutations affecting the C-terminal third of the XPD protein,[41,46] Fujimoto et al described two additional patients with identical mutations affecting the ATPase domain, near the N terminus, on the opposite end of the XPD protein.[40] They also showed the breakdown of DNA in these patients' cells as they had previously shown in the other two cases of XP-D/CS overlap syndrome. Despite their molecular and cellular phenotype, however, these patients differed substantially in their clinical manifestations.

XP-G/CS Overlap Syndrome

Cells derived from individuals who have not only clinical features of CS but also skin changes, including skin cancers, typical of XP have been found to have biochemical features typical of XPG.[47] Nouspikel et al reported that cells from XP-G patients, with severe early onset of CS, produce terminated and unstable XPG proteins and cells from a pair of mildly affected XP-G siblings, without signs or symptoms of CS, are able to synthesize a full length product from one allele with a missense mutation.[48] (Portions of this paper were subsequently withdrawn by the editor, but those portions putting forth the above data and conclusions were defended as correct by the original authors, less the author of the withdrawn part of the paper.[49]) These results were subsequently confirmed by Lalle, et al[50] and by Emmert et al[51] who reported severe XP-G/CS overlap syndrome in two patients with *XPG* mutations expected to lead to severely truncated proteins in both alleles, and only much milder XP-G in a third patient with mutations expected to retain residual activity in one allele. Nouspikel et al concluded that the XP-G protein must have a second function in addition to its role as a structure-specific nuclease of NER.[48,52] XP-G has also been proposed as functioning in transcription coupled base excision repair (TC-BER).

XP-H/CS Overlap Syndrome

Originally only one patient with XP-H/CS was claimed to be suffering from this unique complementation group of XP, but later,[53] following some interchange of publications,[54,55] this group was withdrawn and the mutation was assigned to the XP-D complementation group.[56]

XP/CSB Overlap Syndrome

Cells derived from persons with an XP clinical phenotype and with neurodegenerative changes have been found to have a cellular phenotype typical of CS and a mutation in the *ERCC6/CSB* gene.[57,58] Greenshaw, et al described two siblings with cutaneous photosensitivity and central nervous system dysfunction described as "strongly reminiscent" of the DeSanctis-Cacchione syndrome [XP with severe neurological disease]. Cellular sensitivity to UV radiation was 2-3 times that of controls, but unscheduled DNA synthesis (testing for "classical" XP) following UV radiation was normal and the cells tested negative for the XP variant phenotype. In contrast, recovery of RNA synthesis following UV radiation was markedly delayed, typical of CS.[57] Colella et al reported that these patients had an alteration, determined by sequence analysis, in the CSB gene that was identical to that in the CSB gene of a third patient with clinical features of the classical form of CS.[58] Those authors concluded that alterations in the CSB gene give rise to the same repair defect at the cellular level, but that other genetic and/or environmental factors determine the pathological phenotype.[58]

XP/Progeroid Overlap Syndromes

TTD and CS are both examples of "progeroid syndromes", diseases that resemble, but do not precisely equate with accelerated ageing. They are also called "segmental" progeroid syndrome, meaning that they are incomplete mimics of the full blown progeroid syndrome. These diseases are also associated with DNA repair defects.[59,60] For example, in a recent study Niedernhofer et al described a boy with signs and symptoms of both XP and a segmental progeroid syndrome.[61] He had a null mutation in *XP-F*. The gene product, XP-F, pairs with another protein, ERCC1, as a heteroduplex. This complex has an endonucleolytic DNA cleavage activity, with specificity for DNA exposed to UV radiation inducing thymine dimers and pyrimidine-pyrimidone 6-4 photoproducts, or DNA interstrand cross-links. It induces single stranded cleavages 5′ to the lesion. The patient developed sunburn following minimal sun exposure. Signs of premature ageing appeared late in childhood before puberty, and he died at 15 years of age. This syndrome was proposed by the authors to involve suppression of the somatotroph axis in a complex interplay.[61] The proposed conclusion was based on work with a mouse model in which the *ERCC1* gene was null mutated. Convincing evidence was presented that the phenotype of this mouse model is similar to this boy's disease.

In another recent study, a patient with a biallelic mutation in *ERCC1* who had signs and symptoms of the COFS syndrome was reported.[62] A severe clinical phenotype, despite mild cellular phenotype, suggested an additional role for ERCC1 beyond that in NER. An *ERCC1* deficient mouse model was also described as having a phenotype resembling COFS syndrome. These papers, published virtually simultaneously, one proposing the $ERCC1^{-/-}$ mouse as a model for a progeroid syndrome[61] and the other proposing the $ERCC1^{-/-}$ mouse as a model for the COFS syndrome,[62] have several authors in common, all of whom are excellent investigators. This underscores the difficulties involved in drawing meaningful conclusions from these overlap syndromes and animal and cellular (including yeast) models that are used to study them. Perhaps one explanation for these problems is involvement of additional genes.[6,7]

Conclusion

Study of these overlap syndromes continues to yield surprises that were totally unpredicted and which would have been widely condemned as absurd had they been proposed as hypothetical possibilities. The finding that an identical mutation in the same gene is associated in one patient with one disease and in other patients with a different one flies in the face of commonly accepted notions about genetics, complementation groups and molecular diagnostics. Currently, when a patient is found to have biallelic mutations in a gene, both of which are known to be associated with a genetic disease, a diagnosis is rendered and investigation stops. In how many of these patients have we missed something critical in a second or even third gene? As more efficient sequencing methods become available, it will be worthwhile to reexamine these patients to look for levels of complexity that may presently be unimaginable.

References

1. Wattendorf DJ, Kraemer KH. Xeroderma pigmentosum. Gene Reviews (accessible online as www.genereview.org).
2. Lehmann AR. DNA repair-deficient diseases, xeroderma pigmentosum, Cockayne syndrome and trichothirodystrophy. Biochemie 2003; 85:1101-1111.
3. Kraemer KH, Patrones NJ, Schiffmann R et al. Xeroderma pigmentosum, trichothdystrophy and Cockayne syndrome: A complex genotype-phenotype relationship. Neuroscience 2007; 145(4):1388-1396.
4. Wood RD. Seven genes for three diseases. Nature 1991; 350:190.
5. Cleaver JE, Thompson LH, Richardson AS et al. A summary of mutations in the UV-sensitive disorders: Xeroderma pigmentosum, Cockayne syndrome and trichothiodystrophy. Hum Mut 1999; 14:9-22.
6. Lambert WC, Lambert MW. Corecessive inheritance: A model for DNA repair, xeroderma pigmentosum and mutagenesis. Mutat Res 1985; 145:227-234.
7. Lambert WC, Lambert MW. Corecessive inheritance: A model for DNA repair and other surveillance genes in higher eukaryotes. Mutat Res 1992; 273:179-192.
8. Lambert MW, Lambert WC. Defects in chromatin-associated DNA repair mechanisms in human genetic disease. Progr Nucl Acids Res Molec Biol 1999; 64:257-310.
9. Lambert WC, Gagna CE, Lambert MW. Xeroderma Pigmentosum: Its overlap with trichothiodystrophy, Cockayne syndrome and other progeroid syndromes. In: Ahmad SI and Hanaoka F, Eds. Molecular

Mechanisms of Xeroderma Pigmentosum. Landes Bioscience and Springer Science + Business Media 2008:128-137.
10. Fujiwara Y, Ichihashi M, Kano Y et al. A new human photosensitive subject with a defect in the recovery of DNA synthesis after ultraviolet light. J Invest Dermatol 1981; 77:256-263.
11. Itoh T, Ono T, Yamaizumi M. A new UV-sensitive syndrome not belonging to any complementation groups of xeroderma pigmentosum or Cockayne syndrome: Siblings showing biochemical characteristics of Cockayne syndrome without typical clinical manifestations. Mutat Res 1994; 314: 233-248.
12. Itoh T, Fujiwara Y, Ono T et al. UV(S) syndrome: A new general category of photosensitive disorders with defective DNA repair, is distinct from xeroderma pigmentosum variant and rodent complementation group I. Am J Hum Genet 1995; 56:1267-1276.
13. Horibata K, Iwamoto Y, Kuraoka I et al. Complete absence of Cockayne syndrome group B gene product gives rise to UV-sensitive syndrome but not Cockayne syndrome. Proc Nat Acad Sci USA 2004; 101:15410-15415.
14. Price VH, Odom RB, Ward WH et al. Trichothiodystrophy: Sulfur-deficient brittle hair as a marker for a neuroectodermal symptom complex. Arch Dermatol 1980; 116:1375-1384.
15. Jorizzo JL, Atherton DJ, Crounse RG et al. Ichthyosis, brittle hair, impaired intelligence, decreased fertility and short stature (IBIDS syndrome). Brit J Dermatol 1982; 106:705-710.
16. Itin PH, Pittelkos MR. Trichothiodystrophy: Review of sulfur-deficient brittle-hair syndromes and association with the ectodermal dysplasias. J Am Acad Dermatol 1990; 22:705-717.
17. Tay CH. Ichyosiform erythroderma, hair shaft abnormalities and mental and growth retardation: A new recessive disorder. Arch Dermatol 1971; 104:4-13.
18. Stary A, Sarasin A. The genetic basis of xeroderma pigmentosum and trichothiodystrophy syndromes. Cancer Surv 1996; 26:155-171.
19. Stefanini M, Vermeulen W, Weeda G et al. A new nucleotide-excision-repair gene associated with the disorder, trichothiodystrophy. Am J Hum Genet 1993; 53:817-821.
20. Lehmann AR, Bootsma D, Clarkson SG et al. Nomenclature of the DNA repair genes. Mutat Res 1994; 315:41-42.
21. Giglia-Mari G, Coin F, Ranish JA et al. A new, tenth subunit of TFIIH is responsible for the DNA repair syndrome, Trichothiodystrophy, group A. Nature Genet 2004; 36:714-719.
22. Vermeulen W, Bergmann E, Auriol J et al. Sublimiting concentration of TFIIH transcription/DNA repair factor causes TTD-A trichothiodystrophy. Nature Genet 2000; 26:307-313.
23. Neilan EG. Cockayne syndrome. Gene Reviews, 2006, Available online at <www.genetests.org>
24. Nance MA, Berry SA. Cockayne syndrome: Review of 140 cases. Am J Med Genet 1992; 42:68-84.
25. Rapin I, Lindenbaum Y, Dickson DW et al. Cockayne syndrome and xeroderma pigmentosum. Neurology 2000; 55:1442-1449.
26. Mallery DL, Tanganelli B, Colella S et al. Molecular analysis of mutations in the CSB (ERCC6) gene in patients with Cockayne syndrome. Am J Hum Genet 1999; 62:77-85 (Erratum, Am J Hum Genet 1999; 64:1491).
27. Van Hoffen A, Natarajan AT, Mayne LN et al. Deficient repair of the transcribed strand of active genes in Cockayne syndrome cells. Nucleic Acids Res 1993; 21:5890-5895.
28. Tantin D. RNA polymerase II elongation complexes containing the Cockayne syndrome group B protein interact with a molecular complex containing the transcription factor II H components xeroderma pigmentosum B and p62. J Biol Chem 1998; 273:27794-27799.
29. Balajee AS, May A, Dianov GL et al. Reduced RNA polymerase II transcription in intact and permealised Cockayne syndrome group B-cells. Proc Nat Acad Sci USA 1997; 94:4306-4311.
30. Van Gool AJ, van der Horst GT, Citterio E et al. Cockayne syndrome: Defective repair of transcription. EMBO J 1997; 16:4155-4162.
31. Lehmann AR. Three complementation groups in Cockayne syndrome. Mutat Res 1982; 106:356.
32. Oh KS, Khan SG, Jaspers NGJ et al. Phenotypic heterogeneity in the XPB DNA helicase gene (ERCC3): Xeroderma pigmentosum without and with Cockayne syndrome. Hum Mutat 2006; 27(11):1092-1103.
33. Winkler GS, Araujo SJ, Fiedler U et al. TFII H with inactive XPD helicase functions in transcription initation but is defective in DNA repair. J Biol Chem 2000; 275:4258-4266.
34. Noojin RO. Xeroderma pigmentosum treated with oral methoxalin. Arch Dermatol 1965; 92:422-423.
35. Robbins JH, Kraemer KH, Lutznem MA et al. Xeroderma pigmentosum: An inherited disease with sun sensitivity, multiple cutaneous neoplasms and abnormal DNA repair. Ann Intern Med 1974; 80:221-228.
36. Brumback RA, Yoder FW, Andrews AD et al. Normal pressure hydrocephalus: Recognition and relationships to neurological abnormalities in Cockayne's syndrome. Arch Neurol 1978; 35:337-345.

37. Scott RJ, Itin P, Kleijem WJ et al. Xeroderma pigmentosum-Cockayne syndrome complex in two patients; Absence of skin tumors despite severe deficiency of DNA excision repair. J Am Acad Dermatol 1993; 29:883-889.
38. Vermeulen W, Scott RJ, Rodgers S et al. Clinical heterogeneity within xeroderma pigmentosum associated with mutations in the DNA repair and transcription gene ERCC3. Am J Hum Genet 1994; 54:191-200.
39. Bartenjev I, Butina MR, Potocnick M. Rare case of Cockayne syndrome with xeroderma pigmentosum. Acta Derm Venereol 2000; 80:213-214.
40. Fujimoto M, Leech SN, Theron T et al. Two new patients compound heterozygous for the same mutation demonstrate diverse clinical features. J Invest Dermatol 2005; 125:86-92.
41. Broughton BC, Thompson AF, Harcourt SA et al. Molecular and cellular analysis of the DNA repair defect in a patient in xeroderma pigmentosum complementation group D who has the clinical features of Xeroderma pigmentosum and Cockayne syndrome. Am J Hum Genet 1995; 56:167-174.
42. Lafforet D, Dupuy JA. Xeroderma pigmentosum and Cockayne syndrome Pediatrics. 1978 Apr;61(4):675-676.
43. Van Hoffen A, Kalle WH, de Jong-Versteeg A et al. Cells from XP-D and XP-D-CS patients exhibit equally inefficient repair of UV-induced damage in transcribed genes but different capacity to recover UV-inhibited transcription. Nucleic Acids Res 1999; 27:2898-2904.
44. Graham JR Jr, Greenberg CR, Anyane-Yeboa K et al. Cranio-oculo-facial-skeletal syndrome caused by defective nucleotide excision repair. Am J Hum Genet 1998; 63:A33.
45. Berneburg M, Lowe JE, Nardo T et al. Ultraviolet damage causes uncontrolled DNA breaks in cells from patients with combined features of XP-D and Cockayne syndrome. EMBO J 2000; 19:1157-1166.
46. Takayma K, Salazar EP, Lehmann A et al. Defects in the DNA and transcription gene ERCC2 in the cancer-prone disorder xeroderma pigmentosum group D. Cancer Res 1995; 55:5656-5663.
47. Okinaka RT, Perez-Castro AV, Sena A et al. Heritable genetic alterations in a xeroderma pigmentosum group G/Cockayne syndrome pedigree. Mutat Res 1997; 385:107-114.
48. Nouspikel T, Lalle P, Leadon SA et al. A common mutational pattern in Cockayne syndrome patients from xeroderma pigmentosum group G: implications for a second XPG function. Proc Nat Acad Sci USA 1997; 94:3116-3121.
49. Snyder SH. A common mutational pattern in Cockayne syndrome patients from xeroderma pigmentosum group G: Implications for a second XPG function.Proc Natl Acad Sci USA. 2006;103(51):19606.
50. Lalle P, Nouspikel T, Constantinou A et al. The founding members of xeroderma pigmentosum group G produce XPG protein with severely impaired endonuclease activity. J Invest Dermatol 2002; 118:344-351.
51. Emmert S, Slor H, Busch DB et al. Relationship of neurologic degeneration to genotype in three xeroderma pigmentosum group G patients. J Invest Dermatol 2002;972-982.
52. Clarkson SG. The XPG story. Biochemie 2003; 85:1113-1121.
53. Moshell AN, Ganges MB, Lutzner MA et al. A new patient with both xeroderma pigmentosum and Cockayne syndrome establishes the new xeroderma pigmentosum complementation group H. In: Friedberg EC, Bridges BA, Eds. Cellular Responses to DNA Damage. New York, Alan R. Liss (Publisher), 1983:209-213.
54. Johnson RT, Elliott GC, Squires S et al. Lack of complementation between xeroderma pigmentosum complementation groups D and H. Hum Genet 1989; 81:203-210.
55. Robbins JH. No lack of complementation for unscheduled DNA synthesis between xeroderma pigmentosum complementation groups D and H. [Letter] Hum Genet 1989; 84:99-100.
56. Robbins JH. Xeroderma pigmentosum complementation group H is withdrawn and assigned to group D. Hum Genet 1991; 88:242.
57. Greenshaw GA, Hebert A, Duke-Woodside ME et al. Xeroderma pigmentosum and Cockayne syndrome: Overlapping clinical and biochemical phenotype. Am J Hum Genet 1992; 50:677-689.
58. Colella S, Nardo T, Botta E et al. Identical mutations in the CSB gene associated with either Cockayne syndrome or the DeSanctis-cacchione syndrome. Hum Mol Genet 2000; 9:1171-1175.
59. Andressoo J-O, Hoijmakers JHJ, Mitchell JR et al. Nucleotide excision repair disorders and the balance between cancer and ageing. Cell cycle 2006; 5:2886-2888.
60. Kirkwood T. Ageing: Too fast by mistake. Nature 2006; 444:1015-1017.
61. Niedernhofer LJ, Garinis GA, Raams A et al. A new progerioid syndrome reveals that genotoxic stress suppresses the somatotroph axis. Nature 2006; 444:1038-1043.
62. Jaspers NGJ, Raams A, Silengo MC et al. First reported patient with human ERCC1 deficiency has cerebro-oculo-facial-skeletal syndrome with a mild defect in nucleotide excision repair and severe developmental failure. Am J Hum Genet 2007; 80:457-466.

Biochemical assays for CS, including assessments of DNA repair function and molecular genetic testing of DNA repair genes, have not yet provided a clear classification of these disorders. Mutations in the same gene can cause phenotypes ranging from mild to severe, whereas very different biochemical characteristics can be present in individuals with quite similar clinical phenotypes. Some individuals with the COFS phenotype have biochemical characteristics of the XP-D complementation group, whereas others have biochemical features typical of the XP-G group, suggesting that Cockayne syndrome and COFS do not completely overlap on a biochemical basis.[7,8] These and other diagnostic dilemmas are discussed in detail elsewhere.[9]

Laboratory Testing for CS

Biochemical Assay of Cellular Phenotype

DNA repair assay. Assays for DNA repair, and the related recovery of RNA synthesis (which is characteristically significantly delayed in CS), following ultraviolet (UV) irradiation are routinely performed on skin fibroblasts but may also be carried out on other cell types, such as lymphoblasts. The most consistent findings in CS cells are: (1) marked hypersensitivity to UV radiation; (2) delayed recovery of RNA synthesis and impaired repair of actively transcribed genes following UV exposure, and (3) defective "transcription-coupled DNA repair (TCR)".[10-12] These assays are currently available in specialized laboratories only.

Molecular assays based on Complementation groups. Cells from most individuals with CS can be divided into two complementation groups based on the proteins that carry out DNA repair defect: the CS WD-repeat protein, ERCC-8, CSA, in Cockayne syndrome-complementation group A (approximately 25% of affected individuals) and the excision repair protein ERCC-6 in Cockayne syndrome-complementation group B (approximately 75% of affected individuals).[13] Complementation group assignment may be inferred directly from results of clinically available molecular assays of these genes. Two genes are known to be associated with CS: excision repair cross-complementing group 6 gene (ERCC6), located on chromosome 10q11. Mutation in this gene leads to CS Type B (CSB).[11,14] Excision repair cross-complementing group 8 gene (ERCC8; CKN1), which is located on chromosome 5. Mutation in this gene leads to CS Type A (CSA).[15]

Over 95% of the pathogenic mutations in ERCC6, detected by sequence analysis, are point mutations that are detectable by sequence analysis. Over 95 percent of these are nonsense or frameshift mutations that are predicted to cause formation of a truncated protein. [11,17-20] In contrast, only about 70% of the pathogenic mutations in ERCC8 are missense, nonsense, or splice-site mutations. The other 30% of these pathogenic mutations are large, partial deletions of ERCC8 that typically escape detection by sequence analysis if they are present either in a heterozygous or in a compound heterozygous state.[15,21,22]

Two clinical applications are currently approved for molecular testing of CS: confirmatory diagnostic testing and carrier testing. Since approximately 75% of CS is caused by mutations in the ERCC6 and the remaining 25% by mutations in the ERCC8 gene, the former is sequenced first in testing a proband. If no mutation is identified in ERCC8, then ERCC6 is sequenced. No laboratory, offering molecular genetic testing for prenatal diagnosis of CS, is listed in the GeneTests Laboratory Directory. However, prenatal testing and pre-implantation genetic diagnosis (PGD) may be available in research clinical laboratories to affected families in which disease-causing mutations have been identified. For laboratories offering custom prenatal testing, see the GeneTest Laboratory Directory, or EGN's review of CS in GeneReviews.[1]

Mutations in ERCC6 have also been identified in individuals with COFS, now recognized to be part of the spectra of CS[19]; individuals with of XP who have a severe neurological phenotype (sometimes referred to as the DeSanctis-Cacchione variant,[23] although the appropriateness of this term has been questioned),[24] and the very rare UV-sensitive syndrome (UVSS, See below).[20] In contrast, no nonCS phenotype has been associated with mutations in ERCC8.

Chapter 8

Molecular Basis and Molecular Diagnosis of Cockayne Syndrome

Edward G. Neilan and W. Clark Lambert*

Introduction

As has been pointed out in the "Preface" of this book, and in references 1-5, Cockayne syndrome (CS) spans a wide spectrum of phenotypes that include: CS Type I, the "classic" form; CS Type II, a more severe form with symptoms present at birth [also known as cerebro-oculo-facio syndrome (COFS) and Pena-Shokeir syndrome Type II]; CS Type III, a milder form; and xeroderma pigmentosum-Cockayne overlap syndrome (XP-CS).

CS Type I is characterized by normal prenatal growth but with the onset of growth and developmental abnormalities in the first two years of life. The disease then is fully manifested, with the height, weight and head circumference all far below the fifth percentile. Progressive impairment of vision, hearing and both central and peripheral nervous system function leads to severe disability. Affected individuals typically die in the first or second decade of life. The more severe CS Type II, also known as "connatal" CS, is characterized by growth failure in utero which is manifested at birth, with little or no postnatal neurological development. Congenital cataracts and/or other structural abnormalities of the eye may be present. Early postnatal contractures of the spine (kyphosis, scoliosis) and joints are observed. Affected children usually die by age seven years. CS Type III is characterized essentially by normal growth and cognitive development followed by late onset of signs of the disease. XP-CS includes, in addition to features typical of CS, facial freckling and xerosis (dry skin) followed by early skin cancers characteristic of XP. Features typical of CS include mental retardation, spasticity, short stature, and hypogonadism. However, XP-CS does not include the facial phenotype of CS, skeletal involvement or CNS demyelination and calcifications. We with others have reviewed these overlap syndromes in chapter 7.

Classic CS is diagnosed by major clinical findings (postnatal growth failure and progressive neurological dysfunction) accompanied by other minor criteria. Some cases are atypical and may require molecular genetic testing. Two genes known to be responsible for most cases of CS: *ERCC6* or *CSB* (approximately 75% of affected individuals) and *ERCC8, CKN1,* or *CSA* (approximately 25% of affected individuals). Sequence analyses for both genes are available.[1] CS is inherited as an autosomal recessive trait. No individual with CS has been reported to reproduce. Once the mutation has been identified in the proband, carrier detection in at-risk family members is available on a clinical basis. Prenatal and pre-implantation testing is not routinely available, but may be available through laboratories offering custom prenatal testing. Formal clinical diagnostic criteria have been proposed only for CS Type I.[6] Because of the progressive nature of the disease, the clinical diagnosis of CS becomes more certain as additional signs and symptoms gradually become manifested as the patient grow up. Laboratory testing can be useful for confirming the suspected clinical diagnosis at any stage of disease progression.

*Corresponding Author: W. Clark Lambert—UMDNJ-New Jersey Medical School, Newark, New Jersey, USA. Email: lamberwc@umdnj.edu

Molecular Mechanisms of Cockayne Syndrome, edited by Shamim I. Ahmad.

Diagnosis of NonClassical CS

Children with "connatal" CS (CS Type II) have symptoms of growth failure at birth, and little or no postnatal neurological development. Congenital cataracts and other structural anomalies of the eyes are present in 30% of cases. Affected individuals have arthrogryposis or early postnatal contractures of the spine (kyphosis, scoliosis) and joints. These children typically die by age seven years. CS Type II overlaps clinically with the COFS, which is also referred to as Pena-Shokeir syndrome Type II. Following the identification of a mutation in ERCC6 in a family with COFS, [19] it was recognized that Cockayne syndrome Type II is the correct diagnosis for individuals previously identified to have COFS or Pena-Shokeir syndrome Type II who have ERCC8 or ERCC6 mutations or who have DNA repair abnormalities characteristic of CS.

The term Cerebro-oculo-facio-skeletal syndrome or COFS and its synonym, Pena-Shokeir syndrome Type II, refer to a group of genetically heterogeneous disorders characterized by congenital neurogenic arthrogryposis (multiple joint contractures), microcephaly, microphthalmia and cataracts. Mutations in at least three genes account for the phenotype; therefore, the term COFS should be reserved for those individuals in whom a more precise genetic diagnosis cannot be made. The original cases of COFS, described by Pena and Shokeir among native Canadian families from Manitoba in 1974,[25] have since been shown to be homozygous for a mutation in ERCC6. Cells from these individuals show the same deficiency of transcription-coupled DNA nucleotide excision repair (TC-NER) as cells from those with CS. It is therefore more appropriate to say that these individuals have CS Type II, rather than COFS. Other persons with COFS have a deficiency in global genomic nucleotide excision repair (GG-NER) resulting from mutations in the XP genes XP-G or XP-D. Yet other individuals diagnosed with COFS have no demonstrable DNA repair defect and presumably are affected because of mutations in as yet unidentified gene(s).[1] Recently, DNA sequencing has confirmed the diagnosis of CS Type III in some individuals who have clinical features associated with CS but whose growth and/or cognition exceed the expected phenotype for CS Type I.[1]

Since the discovery of the genes underlying most cases of CS (i.e., ERCC6/CSB and ERCC8/CSA;CKN1), it has become evident that the distinction between genotype, cellular phenotype, and clinical phenotype is not absolute. XP, a somewhat related DNA repair disorder, includes facial freckling and early skin cancers, neither of which are found in CS. Some patients with XP have neurological disorder and some other phenotype of typical of CS, such as mental retardation, spasticity, short stature, and hypogonadism. However, these do not include skeletal dysplasia, the facial phenotype of CS or CNS demyelination and calcifications. Individuals with XP clinical phenotype have been identified in association with a cellular phenotype more typical of CS and with a mutation in the ERCC6 gene.[23,26] Conversely, individuals with clinical features of CS but with UV induced skin cancers (such as those induced in XP) have been assigned to XPB, XPD, or XPG complementation groups based on their biochemical characteristics (the ability to restore normal function to various DNA repair-deficient cell lines).[27-29] Individuals with other features of CS, but lacking sun sensitivity, have been reported.[17] Also a poor correlation between genotype and phenotype for this group of diseases has been noted.[17] It is not yet clear whether genotype-phenotype correlations exist in Cockayne syndrome. Early reports show no obvious genotype-phenotype correlations for mutations in either *ERCC6* (*CSB*) or *ERCC8* (*CSA*), suggesting that the clinical variability within CS may not be due to mutations in these genes alone. Indeed a null mutation of ERCC6 does not produce CS phenotype; instead the mild UV-sensitive syndrome (UVSS) is noted.[20]

Pathogenesis of CS at Molecular Levels

The proteins encoded by ERCC6 and ERCC8 both play important roles in TC-NER, a DNA repair process that preferentially removes UV-induced pyrimidine dimers (CPD) and 6-4 pyrimidine-pyrimidone (6-4 PP) photoproduct and other transcription-blocking lesions, such as bulky adducts and DNA interstrand cross links, from the transcribed strands of active genes. A deficiency of TC-NER is sufficient to explain the cutaneous photosensitivity of individuals with CS. It is difficult, however, to explain the growth failure and neurodegeneration that typify

CS as well as some cases of XP and their overlap syndromes. In contrast to CS, most individuals with XP have normal growth and nearly normal neurological function, despite having combined deficiencies of both TC-NER and "global genomic nucleotide excision repair" (GG-NER). To explain this paradox, a critical role of the ERCC6 and ERCC8 gene products outside of TC-NER has been suggested; these include an auxiliary function in transcription and/or in nonNER forms of DNA repair. [30,31]

The double strand nucleotide polymorphism (dbsSNP) database lists more than a dozen common single nucleotide polymorphisms (SNPs) (normal allelic variants) that are located within the coding sequence of *ERCC6* (*CSB*) and several within the coding sequence of *ERCC8* (*CSA*). Over 20 different pathogenic mutations have been reported in the ERCC6 gene, most of which are nonsense or frameshift.[11,17-20] Several pathogenic mutations in the ERCC8 gene have also been identified, including nonsense mutations, missense mutations, and large partial deletions.[15,21,22] No single class of mutation seems to predominate. A recent report showed multiple abnormal ERCC8 mRNA splice variants in a single individual with CS, although the mutation(s) responsible could not be detected.[32]

The ERCC6 gene encodes a 1493-amino acid protein, containing at least seven domains conserved in DNA and RNA helicases. This protein appears to enhance the elongation process of transcription by RNA polymerase II, and possibly by RNA polymerases I and III. The ERCC8 gene encodes a 396-amino acid protein of 44 kDa. It is a WD-repeat protein, which interacts with the excision repair protein, ERCC-6 and the p44 protein. P44 is a subunit of TFIIH, an RNA polymerase II transcription factor. Both the ERCC8 and ERCC6 gene products appear to be involved in transcription-coupled DNA repair (TCR),[31,33] and possibly in other processes.[30]

Most pathogenic mutations in ERCC6 are nonsense or frameshift that would be predicted to encode a truncated protein. These mutation spectra suggest that the pathogenic mechanism of CS may not be as simple as a loss of normal function. Horibata et al reported that, in at least one case, a homozygous null mutation of *ERCC6* failed to produce CS, instead displaying only a much milder disorder, the ultraviolet sensitive syndrome (UVSS).[20] CS is one of a group of diseases with defective NER leading primarily to neurodegeneration. How the NER is linked to neurodegeneration is unclear.[34] One possibility is that the pathogenic abnormalities, due to mutation in *ERCC8* vary from missense mutations that alter a single amino acid to large, partial deletions of the gene that remove entire exons.[15,21,22]

References

1. Neilan EG. Cockayne syndrome. GeneReviews, 2006. [e-pub online at www.genetests.org]
2. American Res of Pathology. Xeroderma pigmentosum and Cockayne syndrome: A Multimedia Overview. Copyright 1998.
3. de Boer J, Hoeijmakers JH. Nucleotide excision repair and human syndromes. Carcinogenesis 2000;21:453-460.
4. Gandolfi A, Horoupian D, Rapin I et al. Deafness in Cockayne's syndrome: morphological, morphometric, and quantitative study of the auditory pathway. Ann Neurol 1984;15:135-143.
5. Traboulsi EI, De Becker I, Maumenee IH. Ocular findings in Cockayne syndrome. Am J Ophthalmol 1992;114:579-583.
6. Nance MA, Berry SA. Cockayne syndrome: review of 140 cases. Am J Med Genet 1992;42:68-84.
7. Graham JR Jr, Greenberg CR, Anyane-Yeboa K et al. COFS syndrome caused by defective nucleotide excision repair. Am J Hum Genet 1998;63:A33.
8. Moriwaki S, Stefanini M, Lehmann AR et al. DNA repair and ultraviolet mutagenesis in cells from a new patient with xeroderma pigmentosum group G and cockayne syndrome resemble xeroderma pigmentosum cells. J Invest Dermatol 1996;107:647-653.
9. Rapin I, Lindenbaum Y, Dickson DW, Kraemer KH, Robbins JH. Cockayne syndrome and xeroderma pigmentosum. Neurology 2000;55:1442-1449.
10. Venema J, Mullenders LH, Natarajan AT et al. The genetic defect in Cockayne syndrome is associated with a defect in repair of UV-induced DNA damage in transcriptionally active DNA. Proc Natl Acad Sci U S A 1990;87:4707-4711.
11. Troelstra C, van Gool A, de Wit J et al. ERCC6, a member of a subfamily of putative helicases, is involved in Cockayne's syndrome and preferential repair of active genes. Cell 1992;71:939-953.

12. van Gool AJ, Citterio E, Rademakers S et al. The Cockayne syndrome B protein, involved in transcription-coupled DNA repair, resides in an RNA polymerase II-containing complex. EMBO J 1997;16:5955-5965.
13. Stefanini M, Fawcett H, Botta E et al. Genetic analysis of twenty-two patients with Cockayne syndrome. Hum Genet 1996;97:418-423.
14. Troelstra C, Hesen W, Bootsma D et al. Structure and expression of the excision repair gene ERCC6, involved in the human disorder Cockayne's syndrome group B. Nucleic Acids Res 1993;21:419-426.
15. Henning KA, Li L, Iyer N et al. The Cockayne syndrome group A gene encodes a WD repeat protein that interacts with CSB protein and a subunit of RNA polymerase II TFIIH. Cell 1995;82:555-564.
16. Cleaver JE, Thompson LH, Richardson AS, States JC. A summary of mutations in the UV-sensitive disorders: xeroderma pigmentosum, Cockayne syndrome, and trichothiodystrophy. Hum Mutat 1999;14:9-22.
17. Mallery DL, Tanganelli B, Colella S et al. Molecular analysis of mutations in the CSB (ERCC6) gene in patients with Cockayne syndrome. [published erratum in Am J Hum Genet 1999 May;64(5):1491]. Am J Hum Genet 1998;62:77-85.
18. Colella S, Nardo T, Mallery D et al. Alterations in the CSB gene in three Italian patients with the severe form of Cockayne syndrome (CS) but without clinical photosensitivity. Hum Mol Genet 1999;8:935-941.
19. Meira LB, Graham JM Jr, Greenberg CR et al. Manitoba aboriginal kindred with original cerebro-oculo-facio-skeletal syndrome has a mutation in the Cockayne syndrome group B (CSB) gene. Am J Hum Genet 2000;66:1221-1228.
20. Horibata K, Iwamoto Y, Kuraoka I et al. Complete absence of Cockayne syndrome group B gene product gives rise to UV-sensitive syndrome but not Cockayne syndrome. Proc Natl Acad Sci USA 2004;101:15410-15415.
21. Ren Y, Saijo M, Nakatsu Yet al. Three novel mutations responsible for Cockayne syndrome group A. Genes Genet Syst 2003;78:93-102.
22. Cao H, Williams C, Carter M et al. CKN1 (MIM 216400): Mutations in Cockayne syndrome Type A and a new common polymorphism. J Hum Genet 2004;49:61-63.
23. Colella S, Nardo T, Botta E et al. Identical mutations in the CSB gene associated with either Cockayne syndrome or the DeSanctis-cacchione variant of xeroderma pigmentosum. Hum Mol Genet 2000;9:1171-1175.
24. Lambert WC, Gagna CE, and Centurion SA. Xeroderma Pigmentosum. In: Lebwohl MG, Heymann BJ, Coulson, eds. Treatment of Skin Disease: Comprehensive Therapeutic Strategies, Second Edition. London, Mosby/Harcourt Health Science, Ltd., 2006: 694-698.
25. Pena SD, Shokeir MH. Autosomal recessive cerebro-oculo-facio-skeletal (COFS) syndrome. Clin Genet 1974;5:285-293.
26. Greenhaw GA, Hebert A, Duke-Woodside ME et al. Xeroderma pigmentosum and Cockayne syndrome: overlapping clinical and biochemical phenotypes. Am J Hum Genet 1992;50:677-689.
27. Okinaka RT, Perez-Castro AV, Sena A et al. Heritable genetic alterations in a xeroderma pigmentosum group G/Cockayne syndrome pedigree. Mutat Res 1997;385:107-114.
28. Riou L, Zeng L, Chevallier-Lagente O et al. The relative expression of mutated XPB genes results in xeroderma pigmentosum/Cockayne's syndrome or trichothiodystrophy cellular phenotypes. Hum Mol Genet 1999;8:1125-1133.
29. van Hoffen A, Kalle WH, de Jong-Versteeg A et al. Cells from XP-D and XP-D-CS patients exhibit equally inefficient repair of UV-induced damage in transcribed genes but different capacity to recover UV-inhibited transcription. Nucleic Acids Res 1999;27:2898-2904.
30. de Waard H, de Wit J, Andressoo JO et al. Different effects of CSA and CSB deficiency on sensitivity to oxidative DNA damage. Mol Cell Biol 2004;24:7941-7948.
31. van den Boom V, Citterio E, Hoogstraten D et al. DNA damage stabilizes interaction of CSB with the transcription elongation machinery. J Cell Biol 2004;166:27-36.
32. Komatsu A, Suzuki S, Inagaki T et al. A kindred with Cockayne syndrome caused by multiple splicing variants of the CSA gene. Am J Med Genet A 2004;128:67-71.
33. Licht CL, Stevnsner T, Bohr VA. Cockayne syndrome group B cellular and biochemical functions. Am J Hum Genet 2003;73:1217-1239.
34. Itoh M, Hayashi M, Shioda K et al. Neurodegeneration in hereditary nucleotide repair disorders. Brain Dev 1999;21:326-333.

CHAPTER 9

Animal and Yeast Models of Cockayne Syndrome

Shisheng Li*

Abstract

Cockayne syndrome (CS) is a rare inherited disease characterized by postnatal growth failure, impaired development of the nervous system, sensitivity to UV light and dramatically shortened lifespan. The clinical features of the syndrome are presented in two forms (i) those carrying only CS phenotypes and also (ii) those associated with xeroderma pigmentosum (XP), an inherited disease that is highly prone to skin cancer by UV exposure, as XP/CS. A number of animal (mouse and fruit fly) and yeast (*Saccharomyces cerevisiae* and *Saccharomyces pombe*) models have been created, carrying mutations in genes homologous to those implicated in human CS and combined XP/CS. These models, to some extent, recapitulate the phenotypes of the human CS and combined XP/CS at the clinical, cellular and molecular levels. Studies with these models indicate that the etiological causes for cockayne syndrome are at least two fold: a defect in transcription coupled nucleotide excision repair and a defect that is less well defined.

Introduction

Cockayne syndrome (CS) is a rare inherited disease characterized by postnatal growth failure, impaired development of the nervous system, sensitivity to sunlight (photosensitivity) and dramatically shortened lifespan (average 12.5 years).[1,2] Xeroderma pigmentosum (XP) is another rare inherited disease characterized by extreme photosensitivity of the skin and dramatically increased frequency of skin cancer, especially in sun-exposed areas.[2] Individuals with XP sometimes manifest also the clinical features of CS. Patients with combined XP/CS have some features of XP including skin cancer, but usually the clinical features of CS predominate.[3]

Nucleotide excision repair (NER) is a conserved DNA repair mechanism that repairs a wide range of helix-distorting DNA lesions, including UV induced cyclobutane pyrimidine dimers (CPDs) and 6-4 photoproducts (6-4 PPs).[4] NER is a multistep reaction and requires the coordinated action of about 30 proteins implicated in damage recognition, helix opening, lesion verification, strand incision on both sides of a lesion, removal of an oligonucleotide containing the lesion, gap-filling DNA synthesis and the ligation.[4] A specialized NER pathway, called transcription coupled NER (TC-NER), refers to preferential repair of the transcribed strand of an actively transcribed gene. The genome-wide NER process is termed as global genomic NER (GG-NER), to be distinguished from TC-NER.[4]

Mutations of *CSA* and *CSB* genes (Table 1), which are specifically required for TC-NER, but dispensable for GG-NER, can cause typical CS.[2] On the other hand, seven NER genes, designated *XP-A* to *G* are implicated in typical XP or combined XP/CS.[2] *XPA* and *XPF* are essential for both TC-NER and GG-NER and *XPC* and *XPE* (*DDB2*) are specifically required for GG-NER; these

*Shisheng Li—Department of Comparative Biomedical Sciences, School of Veterinary Medicine, Louisiana State University, Baton Rouge, LA 70803, USA. Email: shli@vetmed.lsu.edu

Molecular Mechanisms of Cockayne Syndrome, edited by Shamim I. Ahmad.

Table 1. Human NER genes implicated in XP, CS and combined XP/CS

Gene	Function of Gene Product	Role in NER TC-NER	GG-NER	Syndrome
XPA	Damage recognition/verification	+	+	XP
XPB	3' → 5' DNA helicase; TFIIH subunit	+	+	XP; XP/CS
XPC	Complex with HHR23B; damage recognition	–	+	XP
XPD	5' → 3' DNA helicase; TFIIH subunit	+	+	XP; XP/CS
XPE	WD repeat protein; component of a cullin-containing ubiquitin E3 ligase complex	–	+	XP
XPF	Subunit of a junction-specific endonuclease (XPF/ERCC1) involved in 5' incision	+	+	XP
XPG	Junction-specific endonuclease involved in 3' incision; stabilizes TFIIH	+	+	XP; XP/CS
CSA	WD repeat protein; component of a cullin-containing ubiquitin E3 ligase complex	+	–	CS
CSB	Swi2/Snf2-like DNA-stimulated ATPase	+	–	CS

Adapted from Friedberg et al 2006;[4] Ito et al 2007;[6] and Oh et al 2006.[62]

are exclusively implicated in XP (Table 1).[4] Mutations in *XPB*, *XPD* and *XPG*, which also play essential roles in both TC-NER and GG-NER, can cause either typical XP or combined XP/CS, depending on the types of mutation. All three XP genes i.e., *XPB*, *XPD* and *XPG*, implicated in the combined XP/CS, encode proteins that are to TFIIH, the protein complex involved in both transcription and NER.[5] XPB and XPD proteins are TFIIH subunits. The XPG protein was recently found to play an important role in stabilizing TFIIH.[6]

Several animal and yeast models have been created harboring mutations in genes homologous to those implicated in human CS and combined XP/CS. These models serve as valuable tools for elucidation of the etiological mechanisms of human CS.

Mouse (*Mus musculus*) Models

Typical CS mouse models, which harbor mutations in the *CSA* and/or *CSB* genes and combined XP/CS mouse models, which bear mutations in the *XPD* and *XPG* genes, have been generated. However a combined XP/CS mouse mode, bearing a mutation in the *XPB* gene, has not yet been reported.

CSA Mutant Mice

CSA mutant mice were generated by insertion of the hygromycin selectable marker in exon 2 of the mouse *CSA* gene, resulting in interruption of the coding sequence at amino acid 42.[7] Like cells from CS-A patients, $Csa^{-/-}$ mouse embryonic fibroblasts: (1) are UV sensitive; (2) show normal UV induced unscheduled DNA synthesis (UDS, which reflects GG-NER activities); (3) are deficient in recovery of RNA synthesis (RRS, which reflects TC-NER activities) after UV exposure; and (4) show a defect in TC-NER of UV induced CPDs (Table 2).[7] These $Csa^{-/-}$ mice exhibit UV sensitivity and pronounced age-dependent loss of retinal photoreceptor cells and fail to show the severe developmental and neurological abnormalities of human CS-A. In contrast to human CS-A patients, which are not cancer-prone,[1] $Csa^{-/-}$ mice develop skin tumors after chronic exposure to UVB.[7] Developmental abnormalities, reminiscent of human CS such as severe postnatal growth failure, neurological defects and extremely shortened lifespan (less than one month), are observed in $Csa^{-/-}$ mice that are also homozygously deficient for XPC, which is specifically required for GG-NER but not for TC-NER.[7]

Table 2. Mouse models for CS or combined XP/CS

Gene(s)	Lifespan	CS Symptoms[a]	Sensitivity to UV	TC-NER	GG-NER	Ref.
$Csa^{-/-}$	≥2 years	Mild	Moderate	No	Normal	7
$Csb^{-/-}$	≥2 years	Mild	Moderate	No	Normal	8
$Csa^{-/-}\ Xpc^{-/-}$	<1 month	Severe	ND[b]	ND	ND	7
$Csb^{-/-}\ Xpa^{-/-}$	<23 days	Severe	Severe	No[c]	No[d]	14,15
$Csb^{-/-}\ Xpc^{-/-}$	<23 days	Severe	Severe	No[c]	No[d]	12,15
$Csb^{-/-}\ Oggl^{-/-}$	≥2 years	Mild	Moderate	ND	ND	11-13
Xpd^{XPCS}	~ 15% shortened[f]	Moderate	Moderate	No[c]	No[e]	17
$Xpd^{XPCS}\ Xpa^{-/-}$	<3 weeks	Severe	Severe	ND	ND	17
$Xpg^{-/-}$	<23 days	Severe	Severe	No	No	19
$Xpg^{D811stop}$	Mostly <30 days	Severe	Severe	No	No	21
$Xpg^{\Delta ex15}\ Xpa^{-/-}$	Mostly <30 days	Severe	Severe	No	No	22

[a]May include some or all of these symptoms: postnatal growth failure, neurological defects, cachexia, kyposis and pigmentary retinopathy. [b]Not determined. [c]Inferred from deficiency in recovery of RNA synthesis (RRS) after UV exposure. [d]Inferred from deficiency in UV induced unscheduled DNA synthesis (UDS). [e]No detectable NER, despite 30% UDS activity. [f]Xpd^{XPCS} mice, 546 ± 22.7 days; control, 641 ± 43.3 days.

CSB Mutant Mice

CSB mutant mice were created with a homozygous mutation (K337stop) which mimics the mutant allele in human CS-B patient (CS1AN) and encodes a truncated CSB protein[8] (Table 2). $Csb^{-/-}$ mice exhibit all the NER characteristics of human CS patients: UV sensitivity, inactivation of TC-NER, unaffected GG-NER and inability to resume RNA synthesis after UV exposure. However, other CS features such as postnatal growth failure and neurological dysfunction are present in mild form. After 6 months of age a significant number of $Csb^{-/-}$ mice start to develop a stereotypic hyperactive behavior known as circling or waltzing, suggestive of the development of deafness.[8] Spontaneous loss of retinal photoreceptor cells with age was also observed in the $Csb^{-/-}$ mouse.[9] Unlike CS humans, $Csb^{-/-}$ mice are prone to UVB-induced skin cancer.[8] However, these mutant mice are not abnormally prone to tumors caused by oral administration of benzo[*a*] pyrene.[10]

The basal (steady-state) levels of oxidative DNA damage, in cells from $Csb^{-/-}$ mice, are not different from those in wild-type mice and do not increase with age.[11] $Csb^{-/-}$ mice, that are also homozygous defective for Oggl (a DNA glycosylase involved in base excision repair of oxidative DNA lesions), have also been created.[11-13] Compared to $Oggl^{-/-}$, $Csb^{-/-}\ Oggl^{-/-}$ mice accumulate with age several fold higher levels of oxidated purine modifications in hepatocytes, spleenocytes and kidney cells. The repair of oxidative DNA lesions is only slightly affected in $Csb^{-/-}$ cells, more compromised in $Oggl^{-/-}$ cells, but virtually absent in $Csb^{-/-}\ Oggl^{-/-}$ cells.[11] Despite the deficiency in repair of oxidative DNA damage, $Csb^{-/-}\ Oggl^{-/-}$ mice show essentially the same phenotype as $Csb^{-/-}$ mice.[12]

Severe developmental abnormalities, similar to the human CS phenotype are observed in $Csb^{-/-}$ mice, that are also homozygously deficient for XPA, which is essential for both GG-NER and TC-NER,[14,15] or XPC which is only required for GG-NER but not for TC-NER.[12,15]

Although clinical differences between human CS-A and CS-B have not yet been found, cells from $Csa^{-/-}$ and $Csb^{-/-}$ mice show different responses to oxidative DNA damage.[16] Embryonic fibroblasts from $Csb^{-/-}$ mice, but not $Csa^{-/-}$ mice, are hypersensitive to γ radiation and to paraquat, known to generate reactive oxygen species in cells. When fed with food containing di(2-ethylhexyl) phthalate, which promotes oxidative DNA damage in the liver, $Csb^{-/-}$ mice, but not $Csa^{-/-}$ mice, lose weight. However, both $Csa^{-/-}$ and $Csb^{-/-}$ embryonic stem cells are slightly sensitive to ionizing radiation.[16] Furthermore, when they were exposed to ionizing radiation, both $Csb^{-/-}$ and $Csa^{-/-}$ mice show an increase in apoptotic photoreceptor cells.[9]

Xpd^{XPCS} Mice

Xpd^{XPCS} mutant mice, which bear a homozygous G602D mutation and mimicking a XP-D/CS patient (XPCS2), were created recently[17] (Table 2). Primary embryonic fibroblasts from Xpd^{XPCS} mice have no detectable RRS, indicating an absolute lack of TC-NER activity. Analogous to human XP-D/CS cells, Xpd^{XPCS} cells retained ~ 30% UDS activity and are hypersensitive to UV irradiation.[17] Spontaneously immortalized embryonic fibroblasts from Xpd^{XPCS} mice are moderately sensitive to γ irradiation and show a defect in reactivation of a reporter plasmid containing oxidative DNA damage, consistent with a deficiency in repair ability of DNA damaged oxidatively.[17]

Xpd^{XPCS} mice are even more predisposed to UV induced skin cancer than those entirely NER-deficient (e.g., $Xpa^{-/-}$). Human XP-D/CS cells are able to incise their DNA as efficiently as normal cells following UV exposure.[18] These incisions are, however, uncoupled from the rest of the NER process, resulting in accumulation of uncontrolled DNA breakage.[18] Similar to human XP-D/CS cells, Xpd^{XPCS} mouse cells appear to have an increased number of UV induced DNA incisions that are not coupled to the rest of the NER process.[17] The unique UV-induced skin cancer predisposition in Xpd^{XPCS} mice may be caused by the failure to remove UV lesions and by the uncoupling of the incision event from the rest of the NER process.[17]

The mean lifespan of Xpd^{XPCS} mice (546 days; SE = 22.7) is significantly (~15%) shorter than that of control mice (641 days; SE = 43.3).[17] Further reduction of DNA repair capacity in Xpd^{XPCS} mice by concomitant inactivation of *Xpa*, severely exacerbates CS features. Xpd^{XPCS} $Xpa^{-/-}$ double mutant mice are born normally but display enhanced progressive postnatal growth failure, neurodysfunction and death within 3 weeks.[17]

XPG Mutant Mice

$Xpg^{-/-}$ mutant mice were generated by insertion of the neomycin selectable marker in exon 3 of the mouse *XPG* gene, resulting in unstable *XPG* transcript, either the intact or truncated form (Table 2).[19] Embryonic fibroblasts from $Xpg^{-/-}$ mice are UV sensitive and are unable to repair UV induced DNA lesions. On the other hand, cells from these mutant mice are not sensitive to X-ray and H_2O_2, indicating that they are proficient in repairing oxidative DNA lesions.[19]

$Xpg^{-/-}$ mice exhibited postnatal growth failure, developmental abnormalities and underwent premature death (<23 days), reminiscent of human XP-G/CS.[19] Histological analysis of the cerebellum revealed severe degeneration of the Purkinje cell layer in these mutant mice.[20]

Homozygous *Xpg* mutant mice missing the C-terminal 183 [deletion of exon 15 ($Xpg^{\Delta ex15}$)] and 360 [mutation of D811 to a stop codon ($Xpg^{D811stop}$] amino acids were also generated (Table 2).[21] Embryonic fibroblasts from $Xpg^{\Delta ex15}$ mice are moderately UV sensitive and exhibit residual NER of UV induced DNA lesions. However, embryonic fibroblasts from $Xpg^{D811stop}$ mice are hypersensitive to UV and are unable to repair UV induced DNA lesions. The $Xpg^{D811stop}$ mice exhibit growth retardation and a short life span (<35 days), but the $Xpg^{\Delta ex15}$ mice do not.[21]

Interestingly, $Xpg^{\Delta ex15}$ mice that are also homozygous defective for *Xpa* ($Xpg^{\Delta ex15}$ $Xpa^{-/-}$) exhibit CS phenotype including growth retardation and most of them died within 30 days after birth.[22] Cells from $Xpg^{\Delta ex15}$ $Xpa^{-/-}$ mice are as highly sensitive to UV as those from $Xpa^{-/-}$ mice, indicating they are completely deficient in NER.[22] However, cells from the double mutant mice are not sensitive to X-rays and H_2O_2, indicating they have normal activity for repairing oxidative DNA lesions.[22]

Human XP-G/CS patients produce truncated and unstable XPG proteins, whereas typical XP-G patients are able to synthesize a full length mutant protein that is non-functional for NER[23]. It was proposed that a second function of XPG, in addition to its role in NER, exists in the CS phenotype.[23] Indeed, the mouse models suggest that the disruption of this second function (deletion of exon 15) causes CS when combined with a NER defect.[22]

Fruit Fly (*Drosophila melanogaster*) Models

Homologues of human NER genes, implicated in combined XP/CS, i.e., *hay* (*XPB*),[24] *XPD*[25] and *mus201* (*XPG*),[26,27] have been identified in *D. melanogaster* too. However, the genome of this organism does not contain apparent homologues of *CSA* and *CSB* genes.[28] Also, there is no indication of TC-NER for both CPDs[29] and 6-4 PPs[30] in the organism. However, flies harboring mutations in the *hay* and *mus210* genes have been generated, although no fly models bearing mutant *XPD* gene has yet been reported.

Hay Mutant Flies

Although most of the *hay* mutant alleles are lethal,[24] flies harboring lethal allelic combinations of *hay* were rescued to viability by controlled expression of a wild-type *hay* cDNA under an inducible heat shock promoter. Heat shock conditions that give marginal rescue result in flies exhibiting *haywire* features reminiscent of human CS symptoms: severe neurological defects, defined behavioral abnormalities and a dramatically shortened lifespan.[24] Flies bearing viable hay alleles as hemizogotes or homozygotes are sensitive to UV.[24]

The viable hay^{nc2} allele has a single amino acid change (R652C).[31] This mutation results in accumulation of a poisonous product, a truncated polypeptide of ~ 68 kD (the wild type protein is ~ 90 kD).[31] A transgene named hay^{XPCS}, which has a guanine insertion at nucleotide position 2293 mimicking the frame shift mutation in one XP-B/CS patient,[32] was created and introduced into the fly genome.[33] The hay^{XPCS} transgene also produces a truncated polypeptide (~ 60 kD), but is unable to rescue the lethality of *hay* homozygous flies. The presence of one copy of the hay^{XPCS} transgene in a hay^{nc2}/hay^{nc2} background decreases its viability dramatically. Transgenic flies carrying one copy of hay^{XPCS}, one copy of hay^{nc2} allele and one copy of the wild-type gene reproduced *haywire* defects reminiscent of human CS symptoms.[33]

Mus201 Mutant Flies

The *mus201* encodes a protein of 1236 amino acids.[27] Two *mus201* mutants, $mus201^{A1}$ and $mus201^{D1}$, carry frameshift mutations that cause truncations after residues 376 and 726 respectively.[27] In theory, these mutations should mimic those that cause XPG truncations in XP-G/CS patients.[23] However, the mutant flies are phenotypically normal in the absence of exogenous DNA damaging agents, although they are defective in NER and hypersensitive to UV radiation.[27]

Saccharomyces cerevisiae Models

Homologues of all human NER genes that are implicated in CS and combined XP/CS, i.e., *RAD26* (*CSB*),[34] *RAD28* (*CSA*),[35] *RAD2* (*XPG*),[36] *RAD3* (*XPD*)[37] and *RAD25/SSL2* (*XPB*)[38,39] have been identified in the yeast. A plethora of these alleles have been isolated and characterized.

Rad26 Mutants

Analogous to human CS-B phenotype, *rad26* null mutants exhibit severely impaired TC-NER,[34] and delayed RRS after exposure to UV.[34,40] Consistent with the phenotype of human CS cells, *rad26* cells are hypermutable following exposure to UV light.[35] Unlike human CS-B-cells, *rad26* null mutants are not sensitive to UV.[34] However, deletion of *RAD26* enhances UV sensitivity of GG-NER deficient *rad7* or *rad16* cells.[35,41,42]

A reduced level of transcription and transcription elongation in some CS-B-cells has been reported.[43,44] Analogous to some human CS-B cells, yeast *rad26* cells exhibit deficiency in transcription elongation. A *rad26* deletion mutant manifests moderately reduced induction of certain genes and exhibits growth defects after forced adaptation to new nutrient conditions that require rapid synthesis of mRNA.[45] Growth and transcription defects in *rad26* cells are more pronounced in the presence of 6-azauracil, which depletes the cellular pools of GTP and UTP and lowers the elongation rate of RNA polymerase II.[45]

Interestingly in yeast, not all TC-NER depends on Rad26. Rpb9, a nonessential subunit of RNA polymerase II, mediates a subpathway of TC-NER that is independent of Rad26.[41,46] Rpb4, another non-essential subunit of RNA polymerase II, suppresses the Rpb9 subpathway while facilitating the Rad26 subpathway.[41] Deletion of *SPT4*, which encodes a protein involved in regulating transcription elongation/repression, alleviates the requirement of Rad26 for TC-NER.[47] Its deletion also reinstates TC-NER in cells lacking both Rad26 and Rpb9.[48]

Although it may exist, a CSB (Rad26 homologue) independent TC-NER mechanism has not been documented in mammalian cells. Furthermore, how Rpb4, Rpb9 and Spt4 may contribute to or regulate TC-NER in mammalian cells is currently unknown. Therefore, it remains to be determined to what extent the yeast *rad26* mutants may mimic human CS-B phenotype.

Rad28 Mutants

Unlike human CS-A cells, the yeast *rad28* null mutants are not sensitive to UV, show normal RRS following UV irradiation and do not exhibit any defect in NER.[35,40] However, deletion of *RAD28* slightly enhances UV sensitivity of GG-NER deficient *rad7* and *rad16* cells.[35] Consistent with the phenotype of human CS cells and like *rad26* cells, *rad28* cells are hypermutable following exposure to UV light.[35]

Rad2 Mutants

Like the human XPG protein, which forms a stable complex with the transcription/DNA repair factor TFIIH,[6] the yeast Rad2 protein is stably associated with TFIIH.[49] Analogous to human XP-G/CS cells, *rad2* null mutant cells are hypersensitive to UV,[36,50,51] and are defective in TC-NER and GG-NER.[52] Consistent with their NER defect, these *rad2* mutant cells are totally defective in RRS after exposure to UV.[40] Transcription, after forced adaptation to new nutrient conditions, is slightly impaired in *rad2* strains.[53] This transcription defect is pronounced in the presence 6-azauracil and is synergistic with *rad26* deletion.[53]

A yeast strain (*rad2C*Δ) was created harboring a C-terminal (209 residues) truncated Rad2 protein, analogous to the C-terminal (261 residues) truncated XPG found in XP-G/CS patient 94RD27.[53] Similar to *rad2* deleted cells, the *rad2C*Δ mutant is sensitive to 6-azauracil and exhibits impairment of transcription.[53]

Rad3 Mutants

Like the human XPD protein, the yeast Rad3 protein is the 5′ → 3′ (relative to the DNA strand it binds to) DNA helicase subunit of TFIIH[54] and is essential for cell viability.[37] Rad3 is composed of 778 amino acids and contains seven conserved helicase motifs (I, Ia, II—VI).[55] A plethora of yeast strains bearing different mutations in the *RAD3* gene have been isolated and characterized.[55-58] These mutants are either extremely sensitive to UV and deficient in NER, or exhibit moderate or no UV sensitivity. Currently known mutations in human XPD protein that cause XP-D/CS are either G602D (located in the helicase motif V) or G675R (located downstream of the helicase motif VI).[59] Interestingly, none of the yeast *rad3* mutant alleles identified so far appear to closely mimic those in the currently known human XP-D/CS cells. Therefore, any link between the yeast *rad3* mutants and human XP-D/CS remains to be found.

Rad25 (ssl2) Mutants

Like the orthologous human XPB protein, yeast Rad25/Ssl2 is the 3′ → 5′ DNA helicase subunit of TFIIH,[54,60] and is essential for cell's viability.[39] Rad25/Ssl2 is composed of 843 amino acids and contains seven conserved DNA helicase motifs (I, Ia, II-VI) that are located between amino acids 383 and 695.[39] The region of Rad25/Ssl2 that is essential for cell's viability is located in the central region (residues 81 to 750) of the protein.[61]

All the XP-B/CS patients reported to date appear to harbor either a C → A transversion mutation in the last intron of the *XPB* gene, resulting in alteration of 41 amino acids at the C terminus,[32,62] or a F99S mutation in the protein.[62,63] The *ssl2-XP* allele, which encodes a shortened polypeptide (lacking 94 amino acids from the C-terminus), mimics the mutant *XPB* gene found in XP-B/CS patients.[38] Yeast *ssl2-XP* cells are extremely sensitive to UV[38] and are completely deficient in GG-NER and TC-NER.[64] Another allele, $rad25_{799am}$ which encodes a protein lacking 45 amino acids from the C-terminus, confers moderate UV sensitivity on yeast.[39]

A yeast *RAD25/SSL2* allele encoding a polypeptide that mimics the human F99S mutant XPB has not been reported. Other yeast *RAD25/SSL2* alleles, such as *ssl2-DEAD* (V490A and H491D), *ssl2-1* (W427L),[38] $rad25\text{-}ts_{24}$ (V552I and E556K),[65] *ssl2-182* (L691I)[66] and *ssl2-ttr* (E556K),[67] have been characterized. A link between these alleles and human XP-B/CS remains to be determined. Among these mutants only *ssl2-ttr*, in which the E556K mutation is located between helicase motifs III and IV of the protein, exhibit defect in nuclear mRNA export.[68] The defect in nuclear mRNA export was suggested to be another factor contributing to the human CS phenotype.[68]

Schizosaccharomyces pombe Models

Homologues of human NER genes, implicated in CS and combined XP/CS, i.e., *rhp26*⁺ (*CSB*),[69] *rad13*⁺ (*XPG*),[70,71] *rhp3*⁺/*rad15*⁺ (*XPD*)[72,73] and *ptr8*⁺ (*XPB*)[68] have been identified in *S. pombe*. The *SPBC577.09* gene of *S. pombe*, which may encodes a microtubule-associated protein, has homology to the human *CSA* and the *S. cerevisiae YTM1* genes.[74] Exact function and a mutant allele of the *SPBC577.09* gene have not been documented.

In addition to NER, in *S. pombe*, an alternative pathway exists to repair UV photoproducts is UV Damage Excision Repair (UVER).[75] The UVER pathway, initiated by the UVDE protein, operates equally well for both strands of DNA.[69,76] On the other hand, NER is faster for the transcribed strand than for the nontranscribed strand of DNA in the fission yeast.[69,76]

Rhp26 Mutants

Analogous to human CS-B-cells, the *S. pombe rhp26* null mutants exhibit deficiency in TC-NER and are moderately sensitive to UV, specially when the UVER pathway is nonfunctional.[69] On the other hand, the *rhp26* mutant cells are not sensitive to oxidative DNA damaging agents, such as H_2O_2.[69] *S. pombe* cells lacking Rhp26, Rhp16 (essential for GG-NER but dispensable for TC-NER[77]) and UVDE are much more UV resistant than those lacking Rad13 (homologue of human XPG) and UVDE, indicating that a Rhp26 independent TC-NER subpathway may also exist in *S. pombe*.[69]

Rad13 Mutants

Like XPG for human cells, Rad13 protein is not essential for the viability of *S. pombe*.[70,71] Analogous to human XP-G/CS cells, *rad13* null mutants are sensitive to UV and exhibit deficiency in NER.[69,76] *S. pombe* strains, on the other hand, harboring C-terminal truncated Rad13, are sensitive to 4-nitroquinoline 1-oxide (UV-mimetic chemical) and resistant to the anti-tumor drug trabectedin.[78] Although they have not been studied, these yeast strains may mimic the cells from human XP-G/CS patients, which generally harbor C-terminal truncated XPG proteins.[23]

Rad15 (Rhp3) Mutants

Like *XPD* in humans, *rad15*⁺ is an essential gene in *S. pombe*.[72,73] Despite a UDS level of ~ 30% of normal, similar to, or greater than that in other XP-D cells, human XP-D/CS cells are considerably more sensitive to UV than XP-D cells.[79] It was found that XP-D/CS cells are able to incise their DNA as efficiently as normal cells in response to UV damage.[18] These incisions, however, are uncoupled from the rest of the NER process, resulting in accumulation of uncontrolled DNA breaks.[18] Mutation G675R, corresponding to an alteration found in some XP-D/CS cells,[79] confers the most UV sensitivity of every *rad15* mutations tested,[80] indicating that the G675R mutation closely mimics the one in human XP-D/CS cells.

Ptr8 Mutants

Like functional *XPB* for human cells, *ptr8*+ is essential for *S. pombe*.[68] A temperature sensitive mutant, *ptr8-1*, which encodes T675I mutated protein (located between helicase motifs III and IV of the Ptr8 protein), is highly sensitive to UV irradiation, as are human XP-B- cells. *ptr8-1* cells have a defect in nuclear mRNA export. Expression of human XPB in *ptr8-1* cells complemented the temperature sensitive phenotype and the mRNA export defects. It was proposed that defective nuclear mRNA export could be another factor contributing to the CS phenotype in XP-B/CS patients.[68]

Conclusions

The creation of animal and yeast models for CS has opened various avenues for investigating the complicated molecular mechanisms of the disease. Studies with these models showed that deficiency in TC-NER is one but not the sole cause for the disease. Possible additional reasons, such as deficiency in base excision repair of oxidative DNA lesions and in transcription/transactivation of certain genes, have been proposed (for recent reviews, see [81,82]). Recent data support the theory that deficiency in transcription/transactivation of certain genes is the additional cause for CS. Further studies on the currently available animal and yeast models and newly created models will be helpful in elucidating the causes of the disease.

Acknowledgements

Support was provided by NIH grant ES012718 from the National Institutue of Environmental Health Sciences.

References

1. Nance MA, Berry SA. Cockayne syndrome: review of 140 cases. Am J Med Genet 1992; 42:68-84.
2. Kraemer KH, Patronas NJ, Schiffmann R et al. Xeroderma pigmentosum, trichothiodystrophy and cockayne syndrome: a complex genotype-phenotype relationship. Neuroscience 2007; 145:1388-1396.
3. Lindenbaum Y, Dickson D, Rosenbaum P et al. Xeroderma pigmentosum/cockayne syndrome complex: first neuropathological study and review of eight other cases. Eur J Paediatr Neurol 2001; 5:225-242.
4. Friedberg EC, Walker GC, Siede W et al. DNA Repair and mutagenesis. Washington D.C.: ASM Press, 2006:317-378; 865-918.
5. Cleaver JE. Splitting hairs—Discovery of a new DNA repair and transcription factor for the human disease trichothiodystrophy. DNA Repair (Amst) 2005; 4:285-287.
6. Ito S, Kuraoka I, Chymkowitch P et al. XPG stabilizes TFIIH, allowing transactivation of nuclear receptors: implications for Cockayne syndrome in XP-G/CS patients. Mol Cell 2007; 26:231-243.
7. van der Horst GT, Meira L, Gorgels TG et al. UVB radiation-induced cancer predisposition in Cockayne syndrome group A (Csa) mutant mice. DNA Repair (Amst) 2002; 1:143-157.
8. van der Horst GT, van Steeg H, Berg RJ et al. Defective transcription-coupled repair in Cockayne syndrome B mice is associated with skin cancer predisposition. Cell 1997; 89:425-435.
9. Gorgels TG, van der Pluijm I, Brandt RM et al. Retinal degeneration and ionizing radiation hypersensitivity in a mouse model for Cockayne syndrome. Mol Cell Biol 2007; 27:1433-1441.
10. Wijnhoven SW, Kool HJ, van Oostrom CT et al. The relationship between benzo[a]pyrene-induced mutagenesis and carcinogenesis in repair-deficient Cockayne syndrome group B mice. Cancer Res 2000; 60:5681-5687.
11. Osterod M, Larsen E, Le Page F et al. A global DNA repair mechanism involving the cockayne syndrome B (CSB) gene product can prevent the in vivo accumulation of endogenous oxidative DNA base damage. Oncogene 2002; 21:8232-8239.
12. Laposa RR, Huang EJ, Cleaver JE. Increased apoptosis, p53 up-regulation and cerebellar neuronal degeneration in repair-deficient Cockayne syndrome mice. Proc Natl Acad Sci USA 2007; 104:1389-1394.
13. Trapp C, Reite K, Klungland A et al. Deficiency of the Cockayne syndrome B (CSB) gene aggravates the genomic instability caused by endogenous oxidative DNA base damage in mice. Oncogene 2007; 26:4044-4048.
14. Murai M, Enokido Y, Inamura N et al. Early postnatal ataxia and abnormal cerebellar development in mice lacking Xeroderma pigmentosum group A and cockayne syndrome group B DNA repair genes. Proc Natl Acad Sci USA 2001; 98:13379-13384.
15. van der Pluijm I, Garinis GA, Brandt RM et al. Impaired genome maintenance suppresses the growth hormone—insulin-like growth factor 1 axis in mice with Cockayne syndrome. PLoS Biol 2007; 5:e2.

16. de Waard H, de Wit J, Andressoo JO et al. Different effects of CSA and CSB deficiency on sensitivity to oxidative DNA damage. Mol Cell Biol 2004; 24:7941-7948.
17. Andressoo JO, Mitchell JR, de Wit J et al. An Xpd mouse model for the combined Xeroderma pigmentosum/Cockayne syndrome exhibiting both cancer predisposition and segmental progeria. Cancer cell 2006; 10:121-132.
18. Berneburg M, Lowe JE, Nardo T et al. UV damage causes uncontrolled DNA breakage in cells from patients with combined features of XP-D and Cockayne syndrome. EMBO J 2000; 19:1157-1166.
19. Harada YN, Shiomi N, Koike M et al. Postnatal growth failure, short life span and early onset of cellular senescence and subsequent immortalization in mice lacking the Xeroderma pigmentosum group G gene. Mol Cell Biol 1999; 19:2366-2372.
20. Sun XZ, Harada YN, Takahashi S et al. Purkinje cell degeneration in mice lacking the Xeroderma pigmentosum group G gene. Journal of neuroscience research 2001; 64:348-354.
21. Shiomi N, Kito S, Oyama M et al. Identification of the XPG region that causes the onset of Cockayne syndrome by using Xpg mutant mice generated by the cDNA-mediated knock-in method. Mol Cell Biol 2004; 24:3712-3719.
22. Shiomi N, Mori M, Kito S et al. Severe growth retardation and short life span of double-mutant mice lacking Xpa and exon 15 of Xpg. DNA Repair (Amst) 2005; 4:351-357.
23. Nouspikel T, Lalle P, Leadon SA et al. A common mutational pattern in Cockayne syndrome patients from Xeroderma pigmentosum group G: Implications for a second XPG function. Proc Natl Acad Sci USA 1997; 94:3116-3121.
24. Mounkes LC, Jones RS, Liang BC et al. A Drosophila model for Xeroderma pigmentosum and Cockayne's syndrome: haywire encodes the fly homolog of ERCC3, a human excision repair gene. Cell 1992; 71:925-937.
25. Reynaud E, Lomeli H, Vazquez M et al. The Drosophila melanogaster homologue of the Xeroderma pigmentosum D gene product is located in euchromatic regions and has a dynamic response to UV light-induced lesions in polytene chromosomes. Mol Biol Cell 1999; 10:1191-1203.
26. Houle JF, Friedberg EC. The Drosophila ortholog of the human XPG gene. Gene 1999; 234:353-360.
27. Sekelsky JJ, Hollis KJ, Eimerl AI et al. Nucleotide excision repair endonuclease genes in Drosophila melanogaster. Mutat Res 2000; 459:219-228.
28. Sekelsky JJ, Brodsky MH, Burtis KC. DNA repair in Drosophila: Insights from the Drosophila genome sequence. J Cell Biol 2000; 150:F31-36.
29. van der Helm PJ, Klink EC, Lohman PH et al. The repair of UV-induced cyclobutane pyrimidine dimers in the individual genes Gart, Notch and white from isolated brain tissue of Drosophila melanogaster. Mutat Res 1997; 383:113-124.
30. de Cock JG, van Hoffen A, Wijnands J et al. Repair of UV-induced (6-4)photoproducts measured in individual genes in the Drosophila embryonic Kc cell line. Nucleic Acids Res 1992; 20:4789-4793.
31. Mounkes LC, Fuller MT. Molecular characterization of mutant alleles of the DNA repair/basal transcription factor haywire/ERCC3 in Drosophila. Genetics 1999; 152:291-297.
32. Weeda G, van Ham RC, Vermeulen W et al. A presumed DNA helicase encoded by ERCC-3 is involved in the human repair disorders Xeroderma pigmentosum and Cockayne's syndrome. Cell 1990; 62:777-791.
33. Merino C, Reynaud E, Vazquez M et al. DNA repair and transcriptional effects of mutations in TFIIH in Drosophila development. Mol Biol Cell 2002; 13:3246-3256.
34. van Gool AJ, Verhage R, Swagemakers SM et al. RAD26, the functional S. cerevisiae homolog of the Cockayne syndrome B gene ERCC6. EMBO J 1994; 13:5361-5369.
35. Bhatia PK, Verhage RA, Brouwer J et al. Molecular cloning and characterization of Saccharomyces cerevisiae RAD28, the yeast homolog of the human Cockayne syndrome A (CSA) gene. J Bacteriol 1996; 178:5977-5988.
36. Naumovski L, Friedberg EC. Saccharomyces cerevisiae RAD2 gene: Isolation, subcloning and partial characterization. Mol Cell Biol 1984; 4:290-295.
37. Naumovski L, Friedberg EC. A DNA repair gene required for the incision of damaged DNA is essential for viability in Saccharomyces cerevisiae. Proc Natl Acad Sci USA 1983; 80:4818-4821.
38. Gulyas KD, Donahue TF. SSL2, a suppressor of a stem-loop mutation in the HIS4 leader encodes the yeast homolog of human ERCC-3. Cell 1992; 69:1031-1042.
39. Park E, Guzder SN, Koken MH et al. RAD25 (SSL2), the yeast homolog of the human Xeroderma pigmentosum group B DNA repair gene, is essential for viability. Proc Natl Acad Sci USA 1992; 89:11416-11420.
40. Reagan MS, Friedberg EC. Recovery of RNA polymerase II synthesis following DNA damage in mutants of Saccharomyces cerevisiae defective in nucleotide excision repair. Nucleic Acids Res 1997; 25:4257-4263.

41. Li S, Smerdon MJ. Rpb4 and Rpb9 mediate subpathways of transcription-coupled DNA repair in Saccharomyces cerevisiae. EMBO J 2002; 21:5921-5929.
42. Verhage RA, van Gool AJ, de Groot N et al. Double mutants of Saccharomyces cerevisiae with alterations in global genome and transcription-coupled repair. Mol Cell Biol 1996; 16:496-502.
43. Balajee AS, May A, Dianov GL et al. Reduced RNA polymerase II transcription in intact and permeabilized Cockayne syndrome group B-cells. Proc Natl Acad Sci USA 1997; 94:4306-4311.
44. Selby CP, Sancar A. Cockayne syndrome group B protein enhances elongation by RNA polymerase II. Proc Natl Acad Sci USA 1997; 94:11205-11209.
45. Lee SK, Yu SL, Prakash L et al. Requirement for yeast RAD26, a homolog of the human CSB gene, in elongation by RNA polymerase II. Mol Cell Biol 2001; 21:8651-8656.
46. Li S, Smerdon MJ. Dissecting transcription-coupled and global genomic repair in the chromatin of yeast GAL1-10 genes. J Biol Chem 2004; 279:14418-14426.
47. Jansen LE, den Dulk H, Brouns RM et al. Spt4 modulates Rad26 requirement in transcription-coupled nucleotide excision repair. EMBO J 2000; 19:6498-6507.
48. Li S, Ding B, Chen R et al. Evidence that the transcription elongation function of Rpb9 is involved in transcription-coupled DNA repair in Saccharomyces cerevisiae. Mol Cell Biol 2006; 26:9430-9441.
49. Habraken Y, Sung P, Prakash S et al. Transcription factor TFIIH and DNA endonuclease Rad2 constitute yeast nucleotide excision repair factor 3: implications for nucleotide excision repair and cockayne syndrome. Proc Natl Acad Sci USA 1996; 93:10718-10722.
50. Higgins DR, Prakash L, Reynolds P et al. Isolation and characterization of the RAD2 gene of Saccharomyces cerevisiae. Gene 1984; 30:121-128.
51. Madura K, Prakash S. Nucleotide sequence, transcript mapping and regulation of the RAD2 gene of Saccharomyces cerevisiae. J Bacteriol 1986; 166:914-923.
52. Terleth C, Schenk P, Poot R et al. Differential repair of UV damage in rad mutants of Saccharomyces cerevisiae: a possible function of G2 arrest upon UV irradiation. Mol Cell Biol 1990; 10:4678-4684.
53. Lee SK, Yu SL, Prakash L et al. Requirement of yeast RAD2, a homolog of human XPG gene, for efficient RNA polymerase II transcription. Implications for Cockayne syndrome. Cell 2002; 109:823-834.
54. Feaver WJ, Svejstrup JQ, Bardwell L et al. Dual roles of a multiprotein complex from S. cerevisiae in transcription and DNA repair. Cell 1993; 75:1379-1387.
55. Song JM, Montelone BA, Siede W et al. Effects of multiple yeast rad3 mutant alleles on UV sensitivity, mutability and mitotic recombination. J Bacteriol 1990; 172:6620-6630.
56. Rudolf J, Makrantoni V, Ingledew WJ et al. The DNA repair helicases XPD and FancJ have essential iron-sulfur domains. Mol Cell 2006; 23:801-808.
57. Sung P, Higgins D, Prakash L et al. Mutation of lysine-48 to arginine in the yeast RAD3 protein abolishes its ATPase and DNA helicase activities but not the ability to bind ATP. EMBO J 1988; 7:3263-3269.
58. Sweder KS, Chun R, Mori T et al. DNA repair deficiencies associated with mutations in genes encoding subunits of transcription initiation factor TFIIH in yeast. Nucleic Acids Res 1996; 24:1540-1546.
59. Lehmann AR. The Xeroderma pigmentosum group D (XPD) gene: One gene, two functions, three diseases. Genes Dev 2001; 15:15-23.
60. Guzder SN, Sung P, Bailly V et al. RAD25 is a DNA helicase required for DNA repair and RNA polymerase II transcription. Nature 1994; 369:578-581.
61. Furuchi T, Takahashi T, Tanaka S et al. Functions of yeast helicase Ssl2p that are essential for viability are also involved in protection from the toxicity of adriamycin. Nucleic Acids Res 2004; 32:2578-2585.
62. Oh KS, Khan SG, Jaspers NG et al. Phenotypic heterogeneity in the XPB DNA helicase gene (ERCC3): Xeroderma pigmentosum without and with Cockayne syndrome. Hum Mutat 2006; 27:1092-1103.
63. Vermeulen W, Scott RJ, Rodgers S et al. Clinical heterogeneity within Xeroderma pigmentosum associated with mutations in the DNA repair and transcription gene ERCC3. Am J Hum Genet 1994; 54:191-200.
64. Sweder KS, Hanawalt PC. The COOH terminus of suppressor of stem loop (SSL2/RAD25) in yeast is essential for overall genomic excision repair and transcription-coupled repair. J Biol Chem 1994; 269:1852-1857.
65. Qiu H, Park E, Prakash L et al. The Saccharomyces cerevisiae DNA repair gene RAD25 is required for transcription by RNA polymerase II. Genes Dev 1993; 7:2161-2171.
66. Flom G, Weekes J, Johnson JL. Novel interaction of the Hsp90 chaperone machine with Ssl2, an essential DNA helicase in Saccharomyces cerevisiae. Curr Genet 2005; 47:368-380.
67. Lee BS, Lichtenstein CP, Faiola B et al. Posttranslational inhibition of Ty1 retrotransposition by nucleotide excision repair/transcription factor TFIIH subunits Ssl2p and Rad3p. Genetics 1998; 148:1743-1761.
68. Mizuki F, Namiki T, Sato H et al. Participation of XPB/Ptr8p, a component of TFIIH, in nucleocytoplasmic transport of mRNA in fission yeast. Genes Cells 2007; 12:35-47.

69. Yasuhira S, Morimyo M, Yasui A. Transcription dependence and the roles of two excision repair pathways for UV damage in fission yeast Schizosaccharomyces pombe. J Biol Chem 1999; 274:26822-26827.
70. Carr AM, Sheldrick KS, Murray JM et al. Evolutionary conservation of excision repair in Schizosaccharomyces pombe: Evidence for a family of sequences related to the Saccharomyces cerevisiae RAD2 gene. Nucleic Acids Res 1993; 21:1345-1349.
71. MacInnes MA, Dickson JA, Hernandez RR et al. Human ERCC5 cDNA-cosmid complementation for excision repair and bipartite amino acid domains conserved with RAD proteins of Saccharomyces cerevisiae and Schizosaccharomyces pombe. Mol Cell Biol 1993; 13:6393-6402.
72. Murray JM, Doe CL, Schenk P et al. Cloning and characterisation of the S. pombe rad15 gene, a homologue to the S. cerevisiae RAD3 and human ERCC2 genes. Nucleic Acids Res 1992; 20:2673-2678.
73. Reynolds PR, Biggar S, Prakash L et al. The Schizosaccharomyces pombe rhp3+ gene required for DNA repair and cell viability is functionally interchangeable with the RAD3 gene of Saccharomyces cerevisiae. Nucleic Acids Res 1992; 20:2327-2334.
74. Aves SJ, Hunt C, Xiang Z et al. The mei3 region of the Schizosaccharomyces pombe genome. Yeast 2002; 19:521-527.
75. McCready SJ, Osman F, Yasui A. Repair of UV damage in the fission yeast Schizosaccharomyces pombe. Mutat Res 2000; 451:197-210.
76. Lombaerts M, Tijsterman M, Brandsma JA et al. Removal of cyclobutane pyrimidine dimers by the UV damage repair and nucleotide excision repair pathways of Schizosaccharomyces pombe at nucleotide resolution. Nucleic Acids Res 1999; 27:2868-2874.
77. Lombaerts M, Peltola PH, Visse R et al. Characterization of the rhp7(+) and rhp16(+) genes in Schizosaccharomyces pombe. Nucleic Acids Res 1999; 27:3410-3416.
78. Herrero AB, Martin-Castellanos C, Marco E et al. Cross-Talk between nucleotide excision and homologous recombination DNA repair pathways in the mechanism of action of antitumor trabectedin. Cancer Res 2006; 66:8155-8162.
79. Broughton BC, Thompson AF, Harcourt SA et al. Molecular and cellular analysis of the DNA repair defect in a patient in Xeroderma pigmentosum complementation group D who has the clinical features of Xeroderma pigmentosum and cockayne syndrome. Am J Hum Genet 1995; 56:167-174.
80. Berneburg M, Clingen PH, Harcourt SA et al. The cancer-free phenotype in Trichothiodystrophy is unrelated to its repair defect. Cancer Res 2000; 60:431-438.
81. Friedberg EC, Wood RD. New insights into the combined Cockayne/Xeroderma pigmentosum complex: human XPG protein can function in transcription factor stability. Mol Cell 2007; 26:162-164.
82. Frosina G. The current evidence for defective repair of oxidatively damaged DNA in Cockayne syndrome. Free Radic Biol Med 2007; 43:165-177.

Index

A

Apoptosis 1, 7-9, 12, 21, 43-48, 60, 65, 66, 68, 74
ATPase activity 7, 11, 21, 25, 27, 37, 38-40, 56, 58, 60, 68, 69, 71-73

B

Basal ganglia calcification 85
Base excision repair (BER) 6, 8, 10, 11, 38-40, 43, 53-60, 68, 69, 82, 86, 99, 104
BRCA1/BARD1 24, 25, 31

C

c-Abl 43, 46-48, 60
Cachectic dwarfism 1, 55
Cdk-activating kinase (CAK) 28-30, 69, 70, 74
Cell cycle 1, 9, 10, 28, 43-45, 47, 48, 82
Cerebro-oculo-facial syndrome (COFS) 1, 84, 85, 87, 91-93
Chromatin structure 1, 9, 21, 22, 27, 37-39, 44, 45
c-jun NH2 kinase (JNK) 43, 45-48
Cockayne syndrome (CS) 1-13, 15, 19, 20-22, 24-31, 37, 43, 44, 47, 48, 53, 55, 56, 59, 60, 65-69, 71-75, 81-86, 91-94, 97-104
Cockayne syndrome group B (CSB) 1, 5-12, 19, 20-28, 30, 31, 37-40, 43-48, 53-60, 66-69, 73, 74, 81-84, 86, 91-94, 97-99, 101-103
COP9 signalosome (CSN) 6, 10, 19, 22-24, 67
CSA 1, 5-11, 19, 20-27, 31, 43, 44, 55, 66-69, 74, 81, 84, 91-94, 97, 98, 101, 103
CSB ATPase 47, 68

D

Demyelination 3, 4, 12, 43, 59, 84, 91, 93
Dental carie 3, 4, 43, 84
DNA damage 1, 6-9, 13, 20-22, 24, 31, 43-48, 53-60, 66, 67, 69, 74, 99, 100
DNA repair 8, 19, 20, 25, 37-39, 43-47, 53, 55, 57, 58, 65, 71, 74, 81, 83, 87, 92-94, 97, 100, 102
DNA topoisomerase 26
Drosophila melanogaster 101

E

ERCC1 6, 20-22, 25, 26, 31, 87, 98
ERCC2/XPD 83
ERCC3/XPB 83, 84
ERK 46

G

Genetic counseling 15
Global genome nucleotide excision repair (GG-NER) 6, 11, 20-22, 24, 26, 27, 29, 65-67, 69, 82, 93, 94, 97-99, 101-103
GTF2H5/TTD-A 83

H

Hay 101
Hearing loss 3, 4, 30, 43, 85
Histone acetylation 38-40
"Horse rider" gait 84

K

Kinase 28, 29, 43, 45-47

M

Microcephaly 1, 3, 13, 31, 59, 85, 93
Mus201 101
Mus musculus 98

N

NEDD4 24, 25
Nerve conduction velocity 85
Neurodevelopmental impairment 1-4
Nuclear matrix 21, 22
Nuclear receptor 27-29, 69, 74
Nucleosome remodeling 38, 40
Nucleotide excision repair (NER) 3, 6-8, 10-13, 19-31, 39, 40, 43-46, 60, 65-69, 71-74, 81-84, 86, 87, 93, 94, 97-104

O

Overlap syndrome 81-87, 91, 94
Oxidative damage 5, 7-9, 11, 53, 54, 58, 59, 69, 74
Oxidative DNA damage 53-60, 69, 99, 100
Oxidative stress 8, 38, 39, 46, 47, 53, 57, 59, 60, 66
8-Oxoguanine 54, 57

P

p38 46
p53 7-12, 21, 27, 43-48
Pena-Shokeir 1, 5, 84, 91, 93
Pena-Shokeir syndrome, Type II 84
Phosphatase 45
Phosphorylation 27-29, 37, 43, 45-47, 69, 71
Photosensitivity 3, 4, 12-14, 19, 20, 30, 31, 55, 83, 84, 86, 93, 97
PIBIDS 14, 15
Premature aging 19, 53, 55, 58
Progeria 65, 74
Progeroid syndrome 31, 43, 81, 82, 86, 87

R

RAD2 11, 101
RAD3 101, 102
RAD25 101, 103
RAD26 101
RAD28 101, 102
Retinal pigmentation 2-4, 59
RNA polymerase 5-11, 20-29, 31, 37, 39, 40, 43, 44, 47, 65, 66, 68, 81, 83, 94, 102
RNA polymerase II (pol II) (RNAP II) 5-11, 20, 21-29, 31, 37, 39, 40, 43, 44, 47, 66-69, 74, 81, 83, 94, 102

S

Saccharomyces cerevisiae 6, 7, 39, 71, 97, 101
Schizosaccharomyces pombe 103
Spectrin 82

T

TFB5 83
TFIIH 5, 6, 10, 11, 13, 20-22, 23, 25-30, 44, 69-72, 74, 81, 83, 94, 98, 102
Transcription 5, 6, 8-14, 19-31, 37-40, 43-47, 53, 55, 57, 59, 60, 65-67, 69, 71-74, 81-84, 86, 92-94, 97, 98, 101, 102, 104
Transcription-coupled repair (TCR) 5-7, 9, 10, 13, 19, 21, 22, 37, 39, 43, 44, 47, 48, 55, 56, 65-69, 74, 83, 84, 92, 94
Trichothiodystrophy (TTD) 3, 10-14, 19, 20, 27, 28, 71, 72, 81-86

U

Ubiquitin ligase 6, 10, 19, 22-25, 27, 31, 67, 68
Ultraviolet sensitive syndrome (UVSS) 3, 5, 11, 12, 19-21, 30, 82, 83, 92-94
Ultraviolet (UV) 1, 3-9, 11, 12, 19-25, 27, 30, 31, 37-40, 43-46, 48, 55, 56, 60, 65-68, 73, 82-87, 92-94, 97-104

W

Wrapping of DNA 37

X

Xeroderma pigmentosum/ Cockayne syndrome (XP/CS) 11, 12, 19, 28, 30, 81, 82, 84-86, 97-99, 101, 103
Xeroderma pigmentosum (XP) 1-6, 10-13, 19, 20, 22, 26-31, 43, 55, 60, 65, 66, 71-74, 81-87, 91-94, 97-104
XPB 5, 6, 10, 11, 13, 14, 20, 21, 28, 30, 44, 66, 69-71, 74, 81-85, 93, 98, 101-104
XPB helicase 28, 69
XPD 5, 6, 10, 11, 13, 14, 20, 21, 28-30, 66, 69-74, 81-86, 93, 98, 101-103
XPD helicase 28, 69, 71, 72
XP-F 31, 87
XPF/ERCC1 6, 21, 25, 26, 31, 98

XPG 5, 6, 10, 11, 13, 20-23, 25-30, 66, 73, 74, 81-84, 86, 92, 93, 98, 100-103
XPG endonuclease 73
XP-H 86

For Product Safety Concerns and Information please contact our EU representative GPSR@taylorandfrancis.com Taylor & Francis Verlag GmbH, Kaufingerstraße 24, 80331 München, Germany

Printed and bound by CPI Group (UK) Ltd, Croydon, CR0 4YY
14/09/2026
02216223-0013